Boris Urbanek

Safe take-off with runway analyses

Anchor Academic
Publishing

Urbanek, Boris: Safe take-off with runway analyses. Hamburg, Diplomica Verlag GmbH 2012

ISBN: 978-3-95489-017-0
Print: Anchor Academic Publishing, an Imprint of Diplomica® Verlag GmbH, Hamburg, 2012

Bibliographical Information of the German National Library:
The German National Library lists this publication in the German National Bibliography. Detailed bibliographic data can be found at: http://dnb.d-nb.de

The digital publication (eBook) of this work with the ISBN 978-3-95489-517-5 can be purchased on the general market or directly from the publisher.

CONTENT

LIST OF FIGURES

LIST OF TABLES

LIST OF ABBREVIATIONS

ACN	Aircraft classification number
AIP	Aeronautical information publication
AFM	Aeroplane flight manual
ASD	Accelerate-stop distance
ASDA	Accelerate-stop distance available
CDL	Configuration deviation list
CWY	Clearway
FCOM	Flight crew operating manual
Ft	Feet
hPa	Hectopascal(s)
IAS	Indicated air speed
IEM	Interpretative/explanatory material
LPC	Less paper cockpit
m	meter(s)
MEL	Minimum equipment list
MLM	Maximum landing mass
MMEL	Master minimum equipment list
MRM	Maximum ramp mass
MSLM	Maximum structural landing mass
MSTOM	Maximum structural take-off mass
MTOM	Maximum take-off mass
OAT	Outside air temperature
OM	Operations manual
PCN	Pavement classification number
QRH	Quick reference handbook
SWY	Stopway
TOD	Take-off distance

TODA	Take-off distance available
TOD_{n-1}	Take-off distance with one engine inoperative
TOM	Take-off mass
TOR	Take-off run
TORA	Take-off run available
TOR_{n-1}	Take-off run with one engine inoperative
v_1	Take-off decision speed
v_2	Take-off safety climb speed with a critical engine failure
v_{ef}	Engine failure speed
v_{LOF}	Lift-off speed
v_{MCA}	Minimum control speed in the air
v_{MCG}	Minimum control speed on ground
v_{MBE}	Brake energy limit speed
v_r	Rotation speed
v_s	Stall speed
v_{SR}	Reference stall speed
v-speeds	Speeds associated with take-off

INTRODUCTION

Mainly world economic crisis has proved that the airline industry operates at precarious marginal profits and all companies in this sector need to pay attention to keep balance between their revenues and costs. Although the take-off and landing represent only a small portion of the total operation of an aircraft, performance of these two phases is considered to be very important due to entirely different reasons (1). Based on effort to become more competitive, the airlines struggle to operate at maximum possible payloads. On the other hand, both the take-off and landing are the most strictly regulated segments of a flight. Therefore, an aircraft performance optimization must take place to ensure safety level required by the regulation and also to allow the most effective way of airline operation.

An objective of this book is to study all relevant factors, which could influence the aircraft performance during the take-off phase of a flight, to describe an optimization principle and to propose a calculation method to obtain the maximum operationally allowable mass of an aircraft. This work focuses to do it with accordance with currently valid regulation of the European Union.

Nowadays, the regulator within the area of the European Union is the European Aviation Safety Agency (EASA) together with the European Commission. The EASA lays down the certification requirements for all aeronautical parts. Considering the large transport aeroplanes, formerly used regulation JAR-25 has been replaced with CS-25 with direct force of law for all EASA members. Binding procedures affecting the aircraft operation are specified in EU-OPS commended by the European Commission, but it is estimated that by the end of 2012 this regulation will be also transferred under the responsibility of the EASA (2).

As the optimization to obtain the maximum masses for take-off is forced mainly by economic reasons, this work covers only large commercial airplanes, which are assigned in a performance class A. Into this class, according to EU-OPS, belong multi-engine aeroplanes powered by turbo propeller engines with a maximum approved passenger seating configuration of more than 9 or a maximum take-off mass exceeding 5 700 kg, and all multi-engine turbojet powered aeroplanes.

1 DEFINITION AND PURPOSE OF RUNWAY ANALYSIS

One of the most essential responsibilities of an aircraft operator is to ensure a safe operation according to the authority regulations as well as manufacturer's instructions. Considering the fact that the take-off and landing are the most demanding and dangerous phases of a flight, the operator is specially required to pay attention that all pre-flight calculations for these phases are fulfilled. Moreover, the output data of these calculations need to be perfectly understood by all relevant members of a flight crew, who directly conduct the flight, and a flight dispatching, who uses these data during the planning of aircraft activities.

Besides other factors, the Maximum Take-off Mass (MTOM) at a particular aerodrome for specific local and weather conditions is one of the most limiting figures for the aircraft take-off phase. This mass must be reached at latest on the runway at the time of brake release at the start of take-off. According to EU-OPS, an operator shall ensure that the take-off mass does not exceed the maximum take-off mass specified in the Airplane Flight Manual (AFM) for the pressure altitude and the ambient temperature at the aerodrome at which the take-off is to be made (3). Consequently, the lower figure of the Maximum Structural Take-off Mass (MSTOM) and a calculated maximum performance limited mass for the take-off for specific aerodrome conditions is the MTOM. For a safe aircraft operation it is always necessary to consider the worst case scenario and therefore a performance limited mass for the take-off must also include a possibility of critical engine failure at the decision speed v_1. For that reason, in addition to the MTOM, the appropriate take-off speeds including the decision speed need to be determined.

To determine a performance limited mass most of the operators used to use so called aerodrome tables or tables at manuals provided by the aircraft manufacturer (QRH – Quick Reference Handbook). These data though tend to be very conservative and therefore there is a tendency of the aircraft operators to determine a performance limited take-off mass by using software for so called Runway Analyses (4). The aircraft performance data calculation reflects through the whole airline operation and that is why the airlines nowadays either subcontract or use their own tailored software to optimize the necessary take-off performance calculation.

Similar to the MTOM it is necessary to determine the Maximum Landing Mass (MLM) too. An operator shall ensure that the landing mass of the airplane does not exceed the maximum landing mass specified for the altitude and the ambient temperature expected for the estimated time of landing at the destination and alternate aerodrome (3). This mass is the lower figure of a Maximum Structural Landing Mass (MSLM) stated by the manufacturer and a maximum performance limited landing mass calculated for specific conditions at the destination aerodrome according to the AFM. For a determination of the performance limited landing mass the aerodrome tables, the QRH tables or specialized software for runway analysis are used as well (4).

As a conclusion of above mentioned statements it could be defined that the Runway Analysis facilitates the determination of the maximum allowable take-off and landing masses with associated speeds, based on critical engine failure, for specific airport/runway conditions and various airplane configurations. The limitations observed are those specified in the Civil Aviation Authority approved AFM for the airplane. (5)

2 FACTORS AFFECTING MTOM AND V-SPEEDS

The following chapter describes all factors which could affect the determination of the MTOM and associated v-speeds. A strong relationship between them supposes that there cannot be made a sharp line between those factors affecting the MTOM and those affecting the v-speeds. These figures go hand in hand and determination of a proper value of the MTOM for specific conditions without calculation of the associated v-speeds, and vice-versa, is impossible for a safe operation of the aircraft.

For the purpose of this book v-speeds are defined as speeds associated with a take-off phase of a flight. Regarding operational aspect mainly following speeds belong to this group; however there exist many others as well:

- v_1; take-off decision speed - maximum calibrated airspeed established by the operator at which the crew can at latest decide to reject the take-off and it is ensured to stop the aircraft within the limits of the runway
- v_r; rotation speed – calibrated airspeed at which the pilot initiates the rotation during the take-off to reach v_2 at 35 ft height above the runway surface
- v_2; take-off safety speed – minimum climb speed in terms of calibrated airspeed that must be reached at a height of 35 ft above the runway surface in case of engine failure

2.1 MSTOM

A Maximum Structural Take-off Mass is a maximum permissible total aircraft mass at the start of the take-off run specified in the AFM or operations manual (OM), if more restrictive. This mass is determined in accordance with a flight structure resistance criteria and a resistance of landing gear. An aircraft operator must ensure that the take-off mass (TOM) at the brake release never exceeds the MSTOM as it could result in a damage of the aircraft structure. However it could be equal or lower than the Maximum Ramp Mass (MRM), to allow for a fuel used in start-up and taxi.

2.2 The aerodrome runway distances

Runway length, as one of the most important runway characteristics, influences aircraft operation on a particular runway considerably. Annex 14, Volume I, calls for a calculation of declared distances for a runway intended to use by international commercial air transport, and Annex 15 calls for a reporting of declared distances for each direction of the runway in the State Aeronautical Information Publication (AIP) (6). These declared distances for each runway of a particular aerodrome could be found in part Aerodromes of the AIP. Regarding the take-off these declared distances are:

a. **Take-off Run Available** (TORA); the length of runway declared available by appropriate Authority and suitable for the ground run of an aircraft taking off;

b. **Take-off Distance Available** (TODA); the length of the take-off run available plus the length of the clearway (CWY) available, if provided;

c. **Accelerate-stop Distance Available** (ASDA); the length of the take-off run available plus the length of the stopway (SWY), if provided.

The introduction of SWYs and CWYs and the use of displaced thresholds on runways created a need for accurate information regarding various physical distances available and suitable for the aircraft take-off. Following figure illustrates examples of different configurations of declared distances for various layouts of SWYs, CWYs and displaced thresholds at the aerodrome.

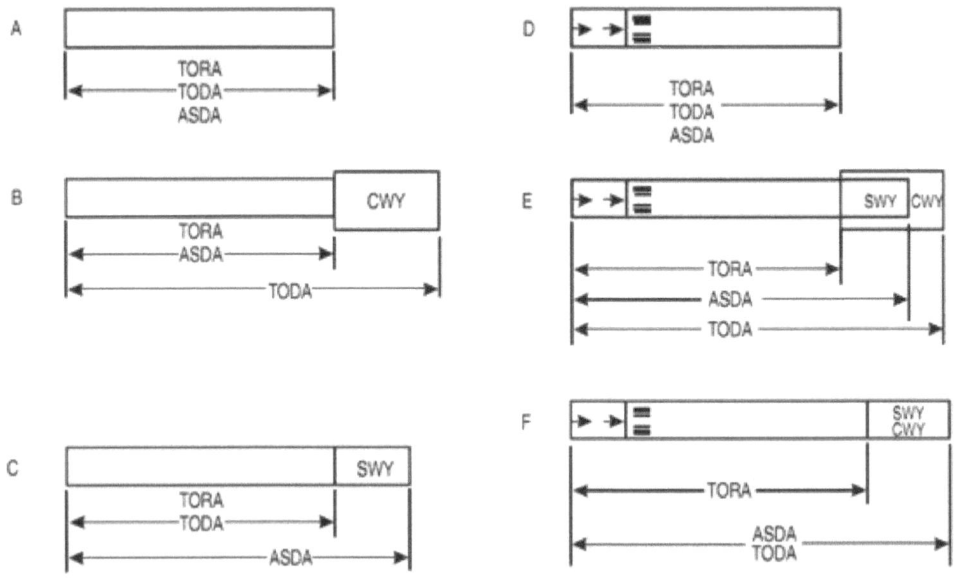

Figure 1: Declared distances [Source: ICAO, Aerodrome design manual, Part 1 Runways]

According to EU-OPS 1.490, an aircraft operator is responsible for proving that distances required by the aircraft for a safe take-off, respectively aborted take-off, do not exceed the appropriate declared distances available at the departure aerodrome. Moreover, when showing compliance with above mentioned, an operator must take into account also the loss, if any, of runway length due to an alignment of the aircraft prior to the take-off.

2.2.1 Line-up distance

Prior to the take-off it is necessary to align the aircraft with runway heading in a direction of intended departure. To utilize all runway length it is favorable to do so at the very beginning of the runway surface dedicated for the take-off. At most aerodromes turns on the runway are required and therefore any time the access to the runway does not permit positioning of the aircraft at the threshold, the line-up corrections are necessary to consider. The alignment distance depends on the aircraft geometry and the access possibility to the runway in use. Usually, it is required 90° turn in order to enter the runway from a taxiway or 180° turnaround on the runway. Typically, the minimum line-up distance is provided by an aircraft manufacturer, but in case of 180° turnaround the minimal runway width, required for performing such turn, has to be additionally considered.

Consequently, the aircraft line-up distance at a particular aerodrome negatively influences the distances of TODA, TORA and ASDA. The reduction of these distances however differs for TODA and TORA on one side and for ASDA on the other. The TODA and TORA adjustments are based on initial distance of the main gear from the beginning of the runway since the screen height is measured from the main gear. On the other hand, adjustment of the ASDA is established upon a distance from the beginning of the runway to the nose gear.

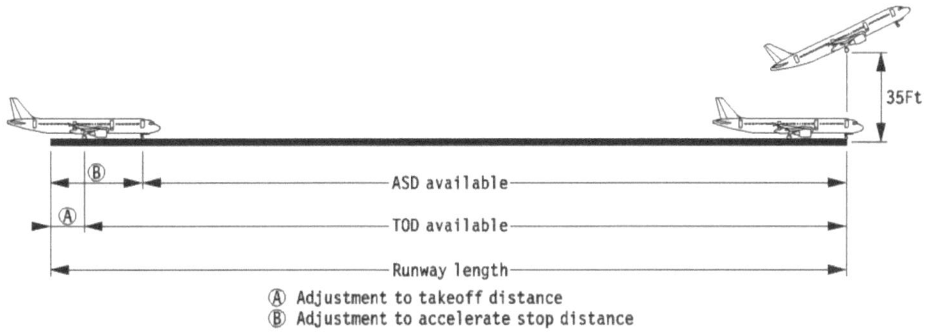

Figure 2: Line-up distance correction [Source: Airbus, Getting to grips of aircraft performance]

2.2.2 Take-off Distance

The Take-off Distance (TOD) of a specific aircraft on a dry runway (for definition see the subchapter Runway condition and contamination) is always greater of:

a. The horizontal distance along take-off path from the start of take-off to the point at which the airplane is 35 ft above the take-off surface assuming that critical engine fails at v_{ef} corresponding to v_1;

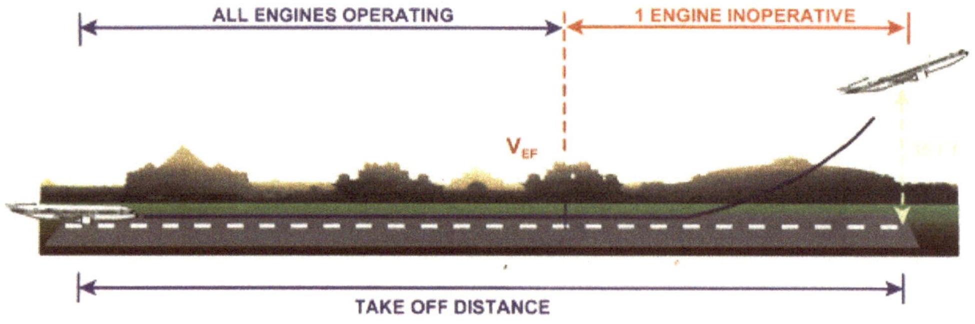

Figure 3: TOD one engine inoperative [Source: Jeppessen, JAA ATPL Training, 032 Performance]

b. 115% of the horizontal distance along the take-off path, with all engines operating, from the start of take-off to the point at which the aircraft is 35 ft above the take-off surface.

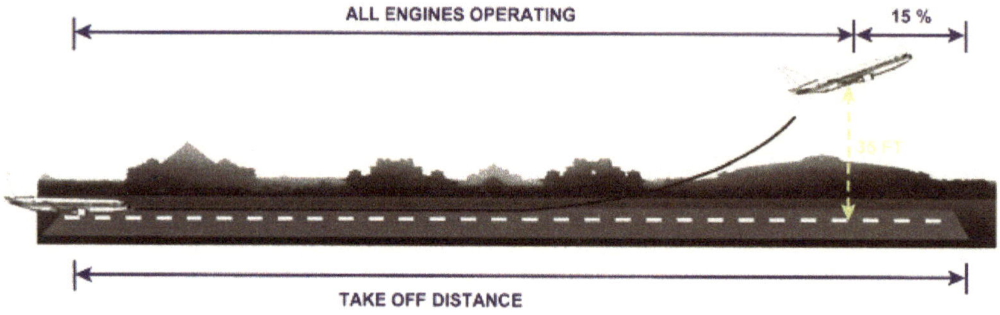

Figure 4: TOD all engines operating [Source: Jeppessen, JAA ATPL Training, 032 Performance]

In case of a wet runway (for definition see the subchapter Runway condition and contamination), the take-off distance is greater of:

a. The take-off distance on a dry runway determined in accordance with the point b. at previous paragraph;

b. The horizontal distance along the take-off path from the start of take-off to the point at which the airplane is 15 ft above the take-off surface, consistent with the achievement of v_2 before reaching 35 ft, assuming that critical engine fails at v_{ef} corresponding to v_1.

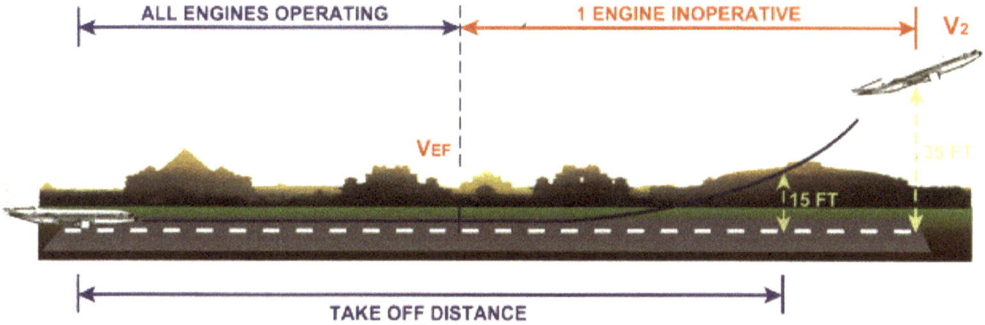

Figure 5: TOD one engine inoperative wet conditions [Source: Jeppessen, JAA ATPL Training, 032 Performance]

2.2.3 Take-off Run

At aerodromes with a runway not provided with clearway the Take-off Run (TOR) of the aircraft is equal to the TOD. If the take-off distance includes a clearway, the take-off run on a dry runway is a figure greater of:

a. The horizontal distance from the start of take-off to a point equidistant between the point at which v_{LOF} is reached and the point at which the airplane is 35 ft above the take-off surface, assuming that the critical engine having failed at $v_{ef;}$

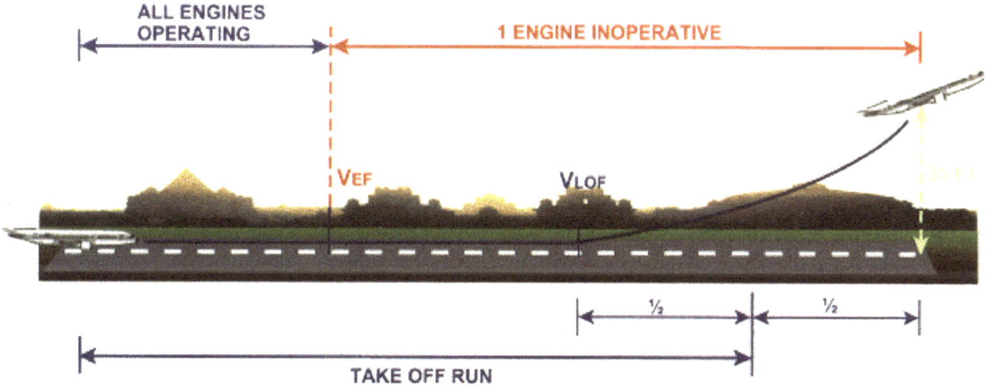

Figure 6: TOR one engine inoperative [Source: Jeppessen, JAA ATPL Training, 032 Performance]

b. 115% of the distance with all engines operating from the start of take-off to a point equidistant between the point at which v_{LOF} is reached and the point at which the airplane is 35 ft above the take-off surface.

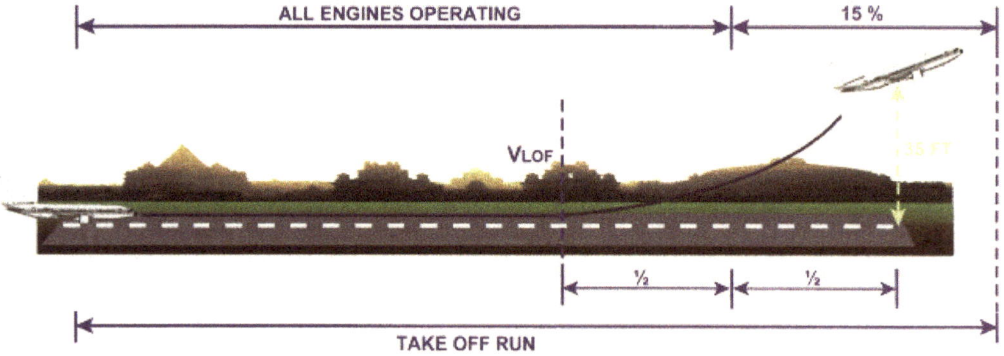

Figure 7: TOR all engines operating [Source: Jeppessen, JAA ATPL Training, 032 Performance]

In case of wet conditions the TOR of the aircraft is defined as greater of:

a. The horizontal distance along the take-off path from the start of take-off to the point at which the airplane is 15 ft above the take-off surface, achieved in a manner consistent with the achievement of v_2 before reaching 35 ft;

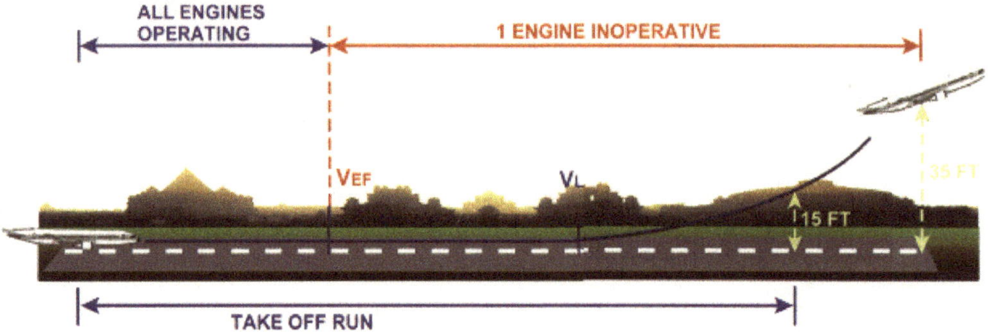

Figure 8: TOR one engine inoperative wet conditions [Source: Jeppessen, JAA ATPL Training, 032 Performance]

b. 115% of the horizontal distance along the take-off path, with all engines operating, from the start of take-off to a point equidistant between the point at which v_{LOF} is reached and the point at which the airplane is 35 ft above the take-off surface (see figure 7).

Both definitions of the TOD and the TOR are specified in CS-25.113.

2.2.4 Accelerate-stop Distance

According to CS-25.109 the Accelerate-stop Distance (ASD) on a dry runway is defined as greater of the following distances:

a. The sum of the distances necessary to:

(i) Accelerate the aircraft from standing start with all engines operating to v_{ef} as for the take-off on a dry runway;

(ii) Allow the airplane to accelerate from v_{ef} to the highest speed reached during the rejected take-off, assuming the critical engine fails at v_{ef} and the pilot takes the first action to reject the take-off at the v_1 for the take-off from a dry runway;

(iii) A distance equivalent to 2 seconds at the v_1 for the take-off from a dry runway;

(iv) Come to a full stop on a dry runway from the point reached at the end of acceleration in (iii) assuming that the pilot does not apply any means of retarding the aircraft until that point is reached.

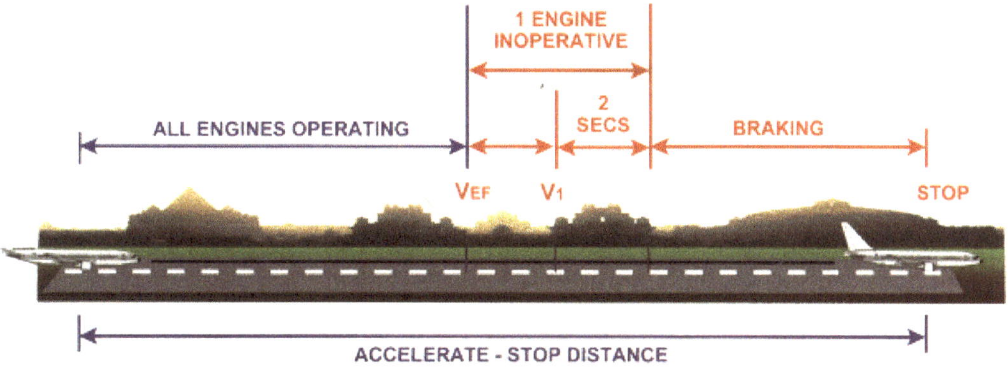

Figure 9: ASD one engine inoperative [Source: Jeppessen, JAA ATPL Training, 032 Performance]

b. The sum of the distances necessary to

(i) Accelerate the airplane from a standing start with all engines operating to the highest speed reached during the rejected take-off, assuming the pilot takes the first action to reject the take-off at the v_1 for the take-off from a dry runway;

(ii) Distance equivalent to 2 seconds at the v_1 for the take-off from a dry runway;

(iii) With all engines still operating, come to a full stop on a dry runway from the speed reached as prescribed in (ii).

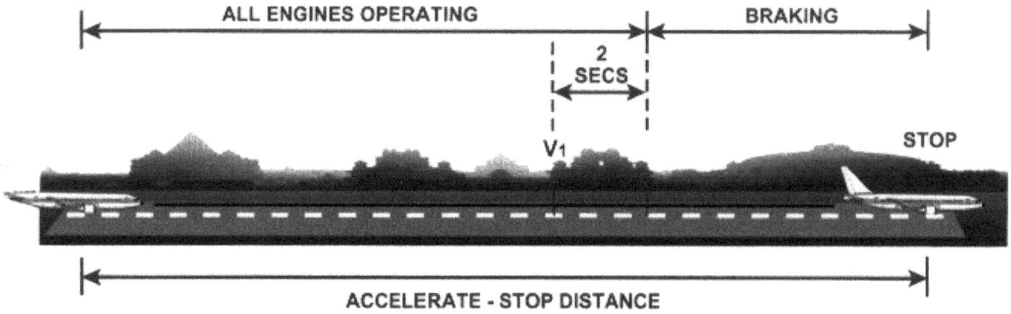

Figure 10: ASD all engines operating [Source: Jeppessen, JAA ATPL Training, 032 Performance]

The accelerate stop distance on a wet runway is the greater of:

a. The accelerate-stop distance on a dry runway determined in accordance with previous paragraphs;

b. The accelerate-stop distance determined in accordance with previous paragraphs, except that the runway is wet and corresponding wet runway values of v_{ef} and v_1 are used.

The values of the TOD, TOR and ASD are directly affected, along with other factors as slope and wind, by mass of the aircraft. The higher the mass is the longer distances of the TOD, TOR and ASD are required by the aircraft, and vice versa. Amount of load influences the aircraft mass and it is how the operator could directly affect the take-off distances required to ensure they would not exceed appropriate values of the TODA, TORA and ASDA at the departure aerodrome.

2.3 Climb limitations

According to the different phase of the take-off flight different climb gradients are required. The take-off path in CS-25.111 is defined as path extending from standing start to a point of the take-off at which the airplane is at height:

- of 1 500 ft above the take-off surface, or
- at which the transition from the take-off to en-route configuration is completed and the final take-off speed is reached,

whichever point is higher. Both definitions assume that aircraft is accelerated on the ground to v_{ef}, at which point the critical engine is made inoperative and remains inoperative for the rest of the take-off, v_2 is reached before the aircraft is 35 ft above the take-off surface and the aircraft continues at this speed minimally until reaching an acceleration height. The take-off flight path is then considered to begin at 35 ft above the take-off surface at the end of the take-off distance.

The take-off flight path can be divided into four segments. Each of these segments is characterized by configuration, thrust setting, speed and requirements for climb gradient so they correspond to the most critical condition prevailing in the segment. The flight path must be based on aircraft's performance without ground effect (7).

Table 1: Take-off segments characteristics

Segment	First	Second	Third	Final
Start when	Reached 35 ft height	Gear fully retracted	Acceleration height reached	En-route configuration achieved
Landing gear	Retraction	Retracted	Retracted	Retracted
Slats/Flaps	Take-off	Take-off	Retraction	Clean
Power	Take-off	Take-off	Take-off	Max continuous
Speed reference	v_2	v_2	Acceleration to final segment speed	Final segment speed

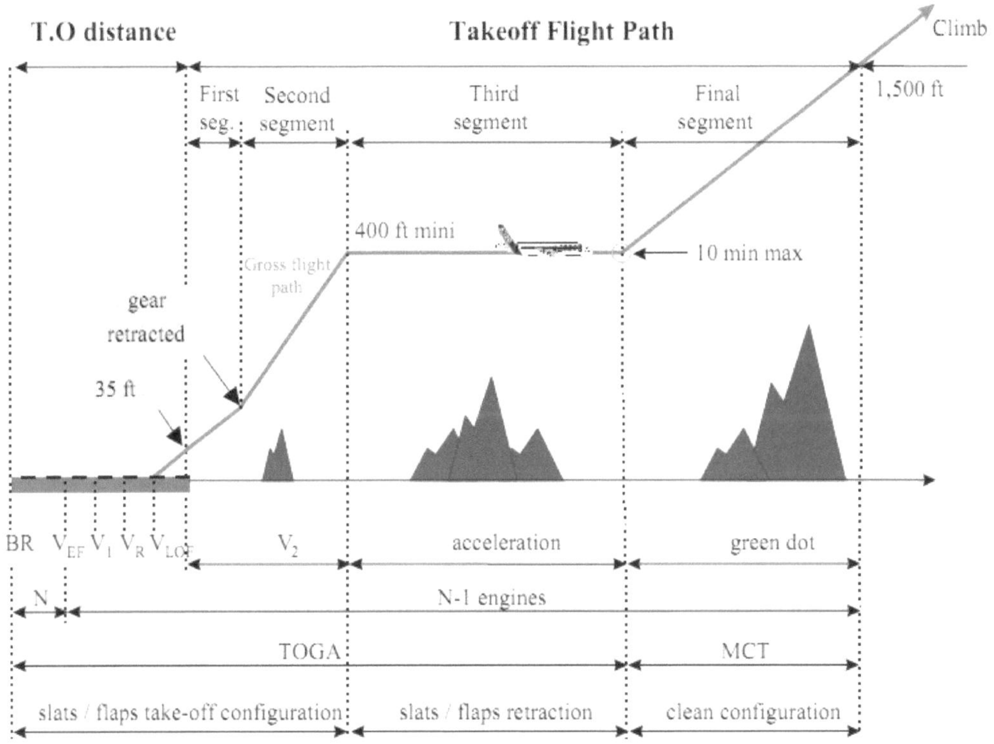

Figure 11: Take-off path segments [Source: Airbus, Getting to grips of aircraft performance]

CS-25.111 states that the airplane must reach v_2 before it is 35 ft above the take-off surface and must continue at a speed as close as practical to but not less than v_2 until it is 400 ft above the take-off surface. Nevertheless, obstacle clearance during the third segment must be ensured at any moment. Therefore, the acceleration height is at least equal or greater than 400 feet. On the other hand, the acceleration height is limited by maximum time of use of maximum take-off thrust, which in case of an engine failure at the take-off is set to 10 minutes. As a result, the en-route configuration at the end of third segment must be achieved within 10 minutes after the take-off.

As mentioned previously, for each segment of the flight there is a minimum climb gradient required by the CS-25.121. These gradients are defined according to the number of engines fitted to the aircraft and they determine a maximum mass for the pressure altitude and ambient temperature at which the aircraft is able to fulfill them. The lowest of these masses determines value limiting MTOM. This is also the reason why in some publications the climb limitation is called WAT, which stands for weight, altitude and temperature. In following table the minimum climb gradient requirements are laid down for each take-off segment.

Table 2: Minimum climb gradient requirements

Segment	Number of engines		
	2	3	4
First	>0	0,3%	0,5%
Second	2,4%	2,7%	3,0%
Final	1,2%	1,5%	1,7%

2.4 Obstacle clearance

Most of the airports are not located in a flatland and they are surrounded by natural or human-made obstacles. Especially during the take-off and landing phase of a flight, when an aircraft is close to the ground, all these obstacles could be possible danger and must be taken into account prior to the take-off. For a safe avoiding of the obstacles, regulation states requirements for both vertical and horizontal plane.

2.4.1 Vertical plane

To ascertain vertical clearance of all relevant obstacles, there was a need to determine a minimum vertical margin between the aircraft and each obstacle. Because of this, gross take-off flight path and net take-off flight path were defined.

The gross take-off flight path is a take-off flight path actually flown by the aircraft from 35 ft above the take-off surface at the end of the take-off distance to the end of the take-off path at optimum conditions (7). This flight path ensures compliance with required climb limitations.

On the other hand, CS-25.115 states that net take-off flight path must be determined so that it represents the actual (gross) take-off flight path reduced at all segments of take-off phase by a gradient equal to:

- 0,8% for two-engined airplanes
- 0,9% for three-engined airplanes
- 1,0% for four-engined airplanes

Therefore, net gradient flight path is gross take-off flight path minus a mandatory reduction. This reduction was introduced due to possible errors at controlling the aircraft by flight crew or some unpredictable conditions during the take-off.

$$Net\ gradient = Gross\ gradient - Gradient\ penalty$$

An operator shall ensure that the net take-off flight path from the 35 ft above runway surface to 1 500 ft above the take-off surface or point at which the transition to en-route configuration is completed, whichever is higher, must clear all relevant obstacles by a vertical distance of at least 35 ft. In case that during any part of the net take-off flight path the airplane is banked by more than 15°, it must clear all obstacles by a vertical distance of at least 50 ft as stated in EU-OPS 1.495. Moreover, the AFM provides a climb gradient decrement depending on aircraft bank turn. The aircraft operator must consider and apply also appropriate amount of this decrement.

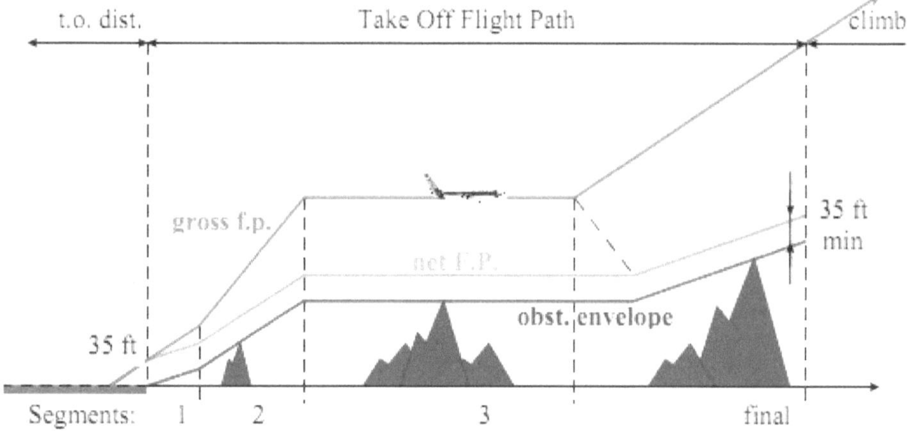

Figure 12: Take-off flight path with obstacles [Source: Airbus, Getting to grips of aircraft performance]

Consequently, obstacles in third segment of the take-off flight path influence minimum acceleration height. This height must be at least 400 ft above take-off surface, but must ensure also the vertical clearance of 35 feet between the net flight path and the obstacles in case of straight take-off and 50 ft if more than 15° bank turn is applied.

2.4.2 Horizontal plane

The vertical margin, as described in previous subchapter, must be kept from all obstacles within surrounding area of the indented track of departure. In some publications this area is also referred as a take-off funnel. Responsibility for clearance of the net take-off flight path from all obstacles within this area lies upon operator. The contour of accountability area is defined in EU-OPS 1.495 and it slightly differs for aircraft with wing span less than 60 meters and those beyond this wing span:

- An aircraft with wingspan greater than 60 meters must strictly clear all obstacles by horizontal distance of at least 90 meters plus 0,125 x D, where D is the horizontal distance the aircraft has traveled from the end of the TODA or the end of the TOD if a turn is scheduled before the end of the TODA.
- For an aircraft with wingspan of less than 60 meters a horizontal obstacle clearance of half the aircraft wingspan plus 60 meters plus 0,125 x D may be used.

From above mentioned it could be observed, that accountability area for obstacle clearance is widening linearly with flown distance of the aircraft. Theoretically, this distance could spread to extensive size, which would be highly unfavorable. This is the reason why a maximum width of surrounding area around the indented departure track was limited.

For those cases where the intended flight path does not require a track change for more than 15°, maximum lateral distance within which it is necessary to consider obstacles is:

- 300 meters, if the pilot is able to maintain the required navigational accuracy through accountability area (defined in AMC-OPS 1.495); or
- 600 meters, for flights under all other conditions.

For those flight paths, were track change of more than 15° is required, an operator needs to consider those obstacles which have horizontal distance less than:

- 600 meters, if the pilot is able to maintain the required navigational accuracy through accountability area (defined in AMC-OPS 1.495); or
- 900 meters for flights under all other conditions.

At each segment of the take-off flight path an aircraft has to clear obstacles at the determined vertical margin. If relevant, the reductions of climb gradient due to bank turn need to be also applied. Taking all obstacles within defined accountability area into account, each of these segments could be influenced by obstacle imposing a minimum net climb gradient. Therefore, a minimum gross climb gradient, higher than one required by the regulation, need to be set. This could result in possible MTOM limitation.

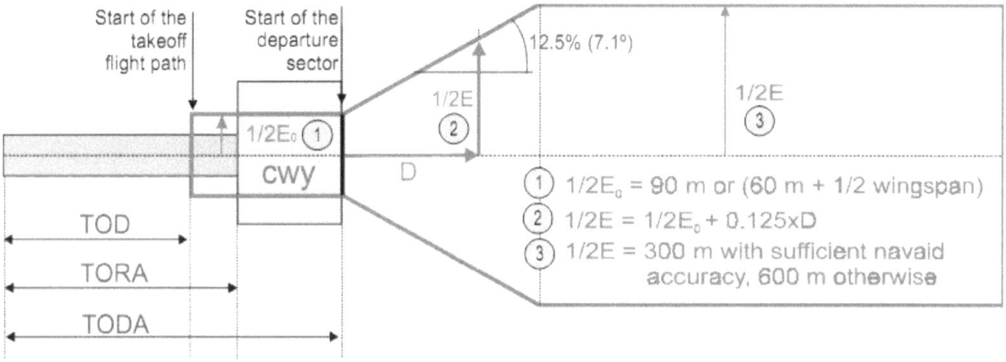

Figure 13: Departure sector (Track change ≤ 15°) [Source: Airbus, Getting to grips of aircraft performance]

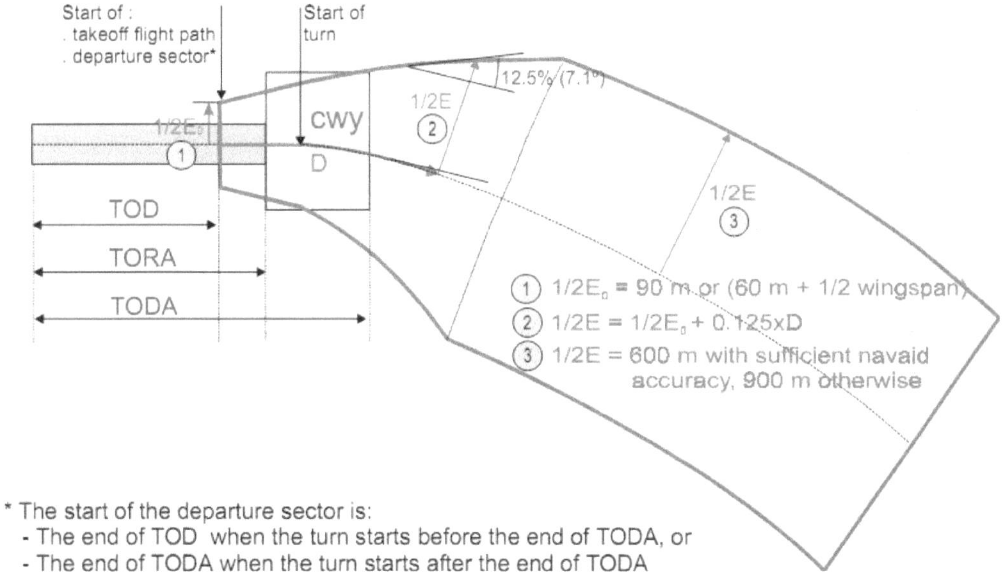

Figure 14: Departure sector (Track change > 15°) [Source: Airbus, Getting to grips of aircraft performance]

2.5 Meteorological elements

Meteorological conditions of a day could significantly influence the determination of the MTOM. Changing of these figures could therefore considerably affect the correct value of the MTOM on a daily or even hourly basis. To this group of affecting elements belong direction and velocity of wind, pressure and temperature of ambient air.

2.5.1 Wind

Considering an influence of a wind on aircraft performance during the take-off phase, it is necessary to calculate a proper value of a wind component along lengthwise axis of the runway resulting in either headwind or tailwind component.

A crosswind component does not affect the value of the MTOM or v-speeds. Nevertheless, to ensure safety of take-off procedures the maximum demonstrated value of cross wind component must be published in the AFM. On the other hand, a wind component along the runway axis affects the take-off ground speeds and consequently required runway distances. If resultant wind component is a headwind, these distances are reduced following possible increasing of the performance limited MTOM. In case of a tailwind it is vice versa.

EU-OPS 1.490 (c) states that an aircraft operator when determining the MTOM must take into account not more than 50 % of the reported head-wind component or not less than 150 % of the reported tailwind component, but these conditions usually form part of the performance tables or software and the operator has to just consider the actual wind component for the MTOM determination.

2.5.2 Pressure altitude

If setting the standard pressure on an aircraft altimeter (1013 hPa), the instrument will read what is known as a pressure altitude. This altimeter setting is used when flying above the transition altitude to comply with a system of flight levels. Moreover, performance curves and tables in the AFM are usually given for varying pressure altitudes rather than aerodrome elevation. This is simply because a change in pressure altitude includes also a change of static pressure and density of the ambient air, which could be different from day to day. In general, the higher the airport pressure altitude the lower static pressure. Because the density of air is proportional to its pressure, the density of ambient air at the aerodrome is proportionally lower too. The consequences of these factors are reduction of a lift for a given true airspeed, reduction of a power and propeller efficiency, if applicable.

$$Lift = \frac{1}{2} \cdot \rho(\downarrow) \cdot S \cdot TAS^2(\uparrow) \cdot c_L$$

The lift equation for a given weight illustrates that if the pressure altitude increases, the true airspeed must be increased to compensate for the lift reduction. Obviously it takes longer to attain the higher forward speed necessary to produce required lift and therefore runway lengths needed for take-off becomes longer. Similarly, a drop in ambient air pressure and density (airport pressure altitude) negatively influences the engine performance as the available take-off thrust is reduced. This reduction could lead to prolonging of the take-off distances and reducing of all take-off climb gradients. Meeting all required climb gradients and available runway distances due to the airport pressure altitude reduction of airframe or engine performance could then result in a limitation of the MTOM.

2.5.3 Temperature

Unlikely to pressure, the density of ambient air is inversely proportional to its temperature. This means if outside air temperature (OAT) raises, the air density decreases. As mentioned in previous subchapter, a drop in air density has consequence in the airframe and engine performance reduction, which must be compensated by increased true air speed. Therefore, the take-off distances required by the aircraft could be prolonged and take-off climb gradients could be impaired, all resulting in a possible MTOM reduction. This is the reason why the operator must take into account the ambient aerodrome temperature at the time of the take-off.

Decreasing ambient air temperature leads consequently to increasing aircraft performance. This is true until possible icing conditions occur. Presence of these conditions depends on several factors, but in general these conditions exist when the OAT is at or below 5°C and sufficient moisture in any form is present (for example clouds, fog, rain, snow, etc.). Reduced lift, increased drag, increased total mass, increased stalling speed and possible block of pressure heads resulting in incorrect reading of the altimeters and speed indicators are just some of ineligible effects of icing. Therefore, the take-off is prohibited when frost, snow or ice is adhering to the wings, control surfaces or propellers and in such case the aircraft need to be de-iced prior to the take-off. Moreover, supplementary data and procedures covering flights in icing conditions must be applied.

2.6 Runway slope

A slope of a runway is usually expressed in percentages, where positive figure is considered to be an upward slope and negative figure downward. A scope of certified slopes is stated in the AFM, but it is usually not smaller than ±2%. In case of required extension to higher limits for operation on a particular runway, additional certification tests are necessary.

From performance point of view, a runway slope influences aircraft acceleration and deceleration ability as well as its maximum brake absorption capability. The ability of the aircraft to stop within the ASDA limited by the brake capability is discussed later in this chapter.

In case of sloping surface, component of weight acts along the runway, influencing the acceleration force. Therefore, an upward slope degrades the aircraft's acceleration ability and increases required take-off distance and take-off run. On the other hand, the stopping distance after rejected take-off is shortened. Depending on a limitation, the MTOM could be then improved or lowered. In case of downhill, it is vice versa, but in general a magnitude of the upward correction is greater than the downward one.

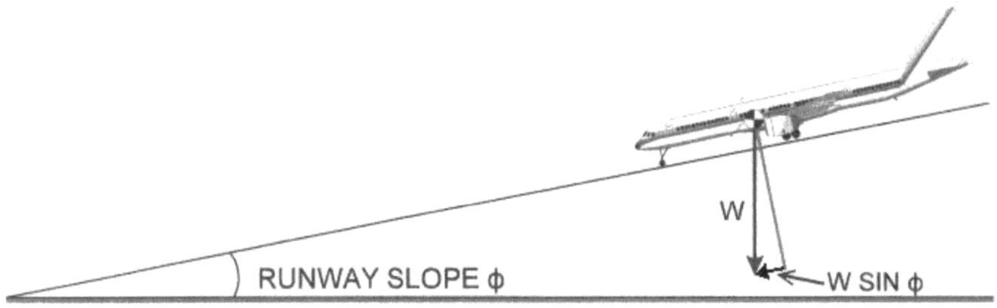

Figure 15: Runway slope [Source: Jeppessen, JAA ATPL Training, 032 Performance]

2.7 Runway condition and contamination

As mentioned in previous, the aircraft operator is responsible to consider a runway state for the MTOM calculation. So far there were discussed different calculation methods of required runway distances for dry and wet runways, but in addition, runway contaminants could also considerably affect the MTOM calculation.

2.7.1 Definitions

In following there are definitions of different runway states as they are stated in EU-OPS 1.480.

Dry runway; a dry runway is one which is neither wet nor contaminated, and includes those paved runways which have been specially prepared with grooves or porous pavement and maintained to retain "effectively dry" braking action even when moisture is present.

Damp runway; a runway is considered damp when the surface is not dry, but when the moisture on it does not give it a shiny appearance.

Wet runway; a runway is considered wet when the runway surface is covered with water, or equivalent, with depth less than or equal to 3 mm, or when there is sufficient moisture on the runway surface to cause it to appear reflective, but without significant areas of standing water.

Contaminated runway; a runway is considered to be contaminated when more than 25% of the runway surface area (whether in isolated areas or not) within the required length and width being used is covered by the following:

i. surface water more than 3 mm deep, or by slush, or loose snow, equivalent to more than 3 mm of water;

ii. snow which has been compressed into a solid mass which resists further compression and will hold together or break into lumps if picked up (compact snow); or

iii. ice, including wet ice (3).

To be able to categorized particular state of a runway it is also necessary to define different contaminants.

Standing water; water of a depth greater than 3mm.

Slush; partly melted snow or ice with a high water content, from which water can readily flow, with an assumed specific gravity of 0,85. Slush is normally a transient condition found only at temperatures close to 0°C.

Wet Snow; snow that will stick together when compressed, but will not readily allow water to flow from it when squeezed, with an assumed specific gravity of 0,5.

Dry Snow; fresh snow that can be blown, or, if compacted by hand, will fall apart upon release (also commonly referred to as loose snow), with an assumed specific gravity of 0,2. The assumption with respect to specific gravity is not applicable to snow which has been subjected to the natural ageing process.

Compacted Snow; snow which has been compressed into a solid mass such that the aircraft wheels, at representative operating pressures and loadings, will run on the surface without causing significant rutting.

Ice; water which has frozen on the runway surface, including the condition where compacted snow transitions to a polished ice surface (8).

2.7.2 Effect on aircraft performance

The contaminant type predicts an effect it has on the aircraft performance. According to the different kind of influence they are divided into hard and fluid. As the hard contaminants reduce only friction forces of the aircraft, the fluid contaminants cause also precipitation drag and aquaplaning. The following table illustrates the categorization of different contaminants.

Table 3: Contaminants categorization

Contaminant	
Hard	**Fluid**
Compacted snow	Water
Ice	Slush
	Loose snow

Reduced friction has always negative effect on the aircraft performance, the TOR is prolonged and in case of rejected take-off the deceleration rate is impaired. On the other hand, it is not true for precipitation drag. It is composed of a displacement drag, produced by the dislodgment of the contaminant fluid from the path of the tire, and a spray impingement drag, produced by the spray thrown up by the wheels onto the fuselage. This additional drag has negative effect for the TOR as acceleration rate is reduced, but following a rejected take-off it has positive effect by improving aircraft deceleration.

The aquaplaning is a phenomenon when water on the runway surface creates an intervening water film between the tire and the runway surface leading to breaking of their contact. Under these circumstances the coefficient of dynamic friction is virtually reduced to zero causing tire traction and the aircraft wheels' braking capacity dropping to almost negligible values. Moreover, the aircraft wheel steering for directional control becomes ineffective.

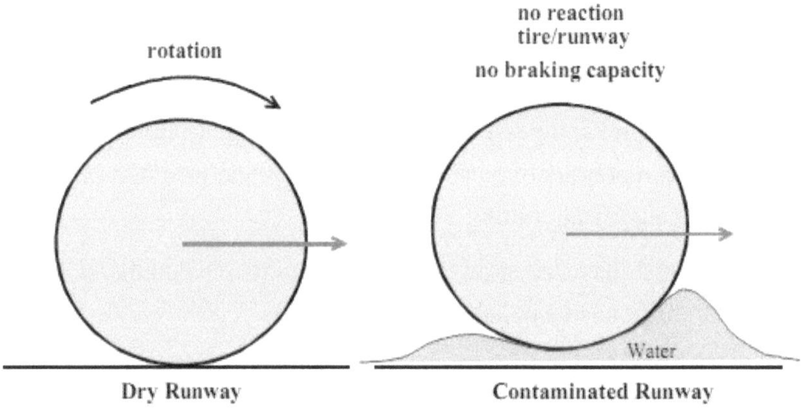

Figure 16: Aquaplaning phenomenon [Source: Airbus, Getting to grips of aircraft performance]

$$v_{AQUQPLANING}[kt] = 34 \cdot (P_T[\frac{kg}{cm^2}]/\sigma)^2$$

The aquaplaning speed, at which the friction forces start to be considerably reduced, depends on a tire pressure and a specific gravity of the contaminant. It is necessary to take into account a hydroplaning penalizing effect when operating on the contaminated runway.

CS 25.1591 clearly states that supplementary performance information furnished by the aircraft manufacturer for runways contaminated with standing water, slush, loose snow, compacted snow or ice must be contained in the AFM. The information may be established by calculation or by testing, but if the information is not supplied, the AFM must contain a statement prohibiting the operation for such contaminant. In addition to performance information, the AFM should also include recommended procedures associated with particular contamination.

When it comes to the TOD, TOR and ASD determination, they are calculated in similar manner as it is on a wet runway although modified by effects of the contaminant. Even more, the ASD calculation on non dry runway surface may take into account also reversers' effect as long as it is available and controllable.

In subchapter Obstacle clearance is stated that the net take-off flight path begins 35 ft above the take-off surface at the end of the TOD and must be guaranteed the vertical margin of 35 ft from all relevant obstacles to the end of the final take-off segment. The screen height (height at the end of the TOD) on a wet or contaminated runway is but reduced to only 15 ft. Consequently, although the net take-off flight path starts at 35 ft, the gross (actual) flight path begins at 15 ft above the take-off surface at the end of TOD. Interpretative/explanatory material (IEM) to EU-OPS 1.495 states that when taking off on a wet or contaminated runway and an engine failure occurs at the point corresponding to the decision speed (v_1) for a wet or contaminated runway, this implies that the airplane can initially be as much as 20 ft below the net take-off flight path and therefore may clear close-in obstacles by only 15 ft (9).

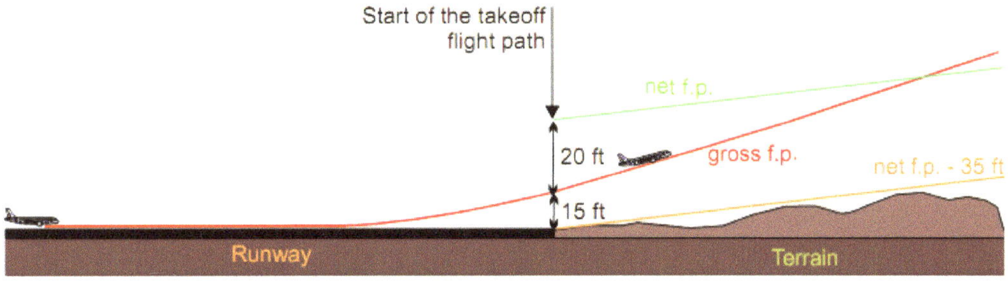

Figure 17: Take-off flight path on a wet and contaminated runway [Source: Airbus, Getting to grips of aircraft performance]

Different requirements for wet and contaminated runways, especially reduced screen height and allowed reverse thrust for the ASD determination in contrast to dry runways, could possibly lead to obtaining lower figures of the TOD and ASD. Depending on limitation it could result in a higher available take-off mass on such runway. This is the reason why the regulation specifies that on a wet or contaminated runway, the MTOM must not exceed the mass, which is permitted for the take-off on a dry runway under the same conditions.

2.8 Tire speed limit

The tire manufacturer specifies a speed limitation called a maximum tire speed (v_{TIRE}). The maximum tire speed is referred as a true ground speed, and so limiting indicated air speed (IAS) and consequent take-off mass varies with the pressure altitude, temperature, wind speed, flaps setting but not with runway slope (22). This speed must prevent from tire structure damage and it is usually verified to allow all weights and associated v-speeds. Considering the fact that this limitation is only relevant until an aircraft is still on ground, v_{LOF} is the most critical value and v_{TIRE} must be always higher for a safe aircraft operation.

$$v_{LOF} \leq v_{TIRE}$$

2.9 Brake energy capacity

In case of aborted take-off before reaching the decision speed v_1 there is a need to stop an aircraft within the accelerate-stop distance available safely. A proportion of the kinetic energy reached at the time of the decision is absorbed by brakes and it is dissipated into the heat. This aircraft kinetic energy at the decision point is proportional to the take-off mass and a square of the decision speed v_1.

$$E_K = \frac{1}{2} \cdot TOM \cdot v_1^2$$

Brakes manufacturers have to specify a maximum brake absorption capacity and so for a given mass there is a limiting speed at which a safe stop could be made. This speed is stated as ground speed and corresponding IAS varies with pressure altitude, temperature and wind. In contrast to tire speed limit, the maximum brake energy speed depends also on a slope of the runway as a change in height involves also a change in potential energy. In CS-25.109 defines: A flight test demonstration of the maximum brake kinetic energy accelerate-stop distance must be conducted with not more than 10 % of the allowable brake wear range remaining on each of the airplane wheel brakes (8). This means that for certification purposes, the absorption capacity must be in addition demonstrated with worn brakes.

The brake energy limit speed v_{MBE} must be determined in compliance with certified maximum brake energy capacity and must be less than or equal to v_1. For most aircraft, v_{MBE} is only limiting in extreme adverse of an altitude, temperature, wind or runway slope and it is verified to allow all masses for usual conditions. If calculated v_1 for a given TOM is higher, the AFM will give the amount of weight to be deducted for each knot that v_1 exceeds v_{MBE} until they are equal.

If the take-off is initiated after prolonged taxiing the brakes could be already at fairly high temperature and their ability to absorb further energy could be impaired. In such cases, the flight manual specifies a time to allow for the brakes to cool.

2.10 Aircraft configuration and systems setting

An aircraft performance always depends on its configuration. It is obvious that gear retraction immediately after reaching the screen height positively contributes to the climb gradients, but when it comes to various configuration settings of flaps and slats it may not be so apparent. While particular configuration gives better performance on a short take-off distance, another gives better climb gradient after the lift-off. Throughout the optimization process it is required to figure out such configuration, which provides with the highest MTOM.

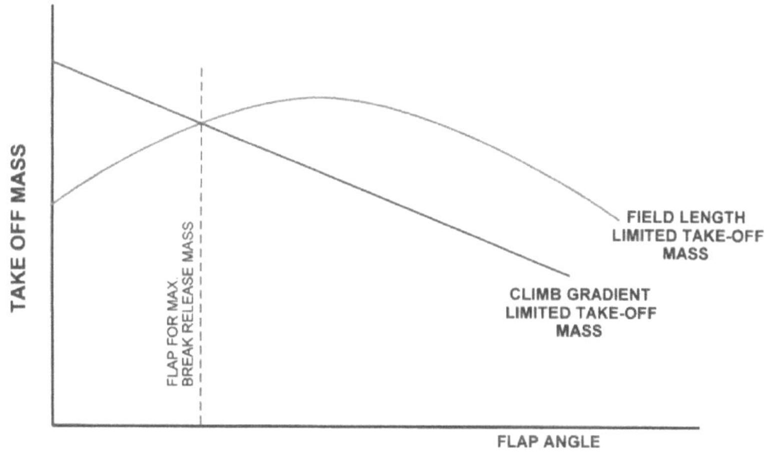

Figure 18: Optimum flap setting [Source: Jeppessen, JAA ATPL Training, 032 Performance]

Moreover, settings of several aircraft systems could also influence the aircraft performance. The meant systems, when switched on, decrease the available engine power and thus degrade the performance. Some of these systems, for example air conditioning, are advised to be turned off for the take-off and their use is only upon the operator discretion, but some of them, for example anti-icing, may be unavoidable for particular aerodrome conditions.

2.11 Aircraft status

Functionality of some systems such as anti-skid, autobrake, propeller feathering, engine reverse, PMC (Power management control), etc., also affects the airplane performance (4). The service of an aircraft with inoperative system is always subjected to the Master Minimum Equipment List (MMEL) or the Minimum Equipment List (MEL) if more restrictive. In case of missing any external parts of the aircraft the conditions under which the aircraft operation could be carried are specified in the Configuration Deviation List (CDL). Moreover, these lists contain procedures to be followed and performance corrections to be applied in such cases. In general, it results in prolonging the required distances for the take-off, impairing the climb gradients and therefore limiting the MTOM.

2.12 Bearing strength

A bearing strength of all aerodrome maneuvering area pavements is expressed by a method of a Pavement Classification Number (PCN), which is compared to an Aircraft Classification Number (ACN). The PCN is a number expressing a bearing strength of a pavement for unrestricted operations considering the type of the pavement, subgrade, tire pressure category and method of a calculation. On the other hand, the ACN is a number expressing a relative effect of an aircraft on the pavement for a specified standard subgrade category. For the calculation of the actual ACN for arbitrary mass between the maximum and empty mass of an airplane, a linear function is applied (10). The aircraft operation on the pavement is permissible, only if the ACN is less or equal to the reported PCN. If higher, an equivalent mass must be deducted from the total aircraft mass to meet at least the highest PCN for the specific pavement.

Under certain conditions it is possible to exceed the calculated load. However, it must not be a regular occurrence and the ACN of considered aircraft should not exceed the reported PCN for more than 10 % for flexible pavements and 5 % for rigid pavements. Moreover, the annual number of overload movements should not go over 5 % of the total annual aircraft movements.

3 TAKE-OFF DATA OPTIMIZATION PRINCIPLE

After considering all influencing factors, it is necessary to set a value of the TOM, which complies with all of them. It is not a hard task to select a specific TOM and prove it does meet all requirements, but the effort to operate at maximum possible payloads calls for optimized values. These provide with maximum values of the TOM and meet regulation and safety requirements too. This section explains a principle of the optimization process.

The factors affecting the values of the MTOM and v-speeds are described in previous chapter, but first of all when considering these parameters in the optimization process it is necessary to determine which of those are fixed and which of them are free to choice. The sustained parameters, for instance the OAT, cannot be changed, but on the other hand a change of free parameters represents a source of the optimization to determine highest possible MTOM. In other words, at given sustained parameters, the take-off mass limitations are set as a function of free parameters. The table below lists influencing parameters in mentioned categories.

Table 4: Parameters influencing take-off

<table>
<tr><th colspan="3">Take-off parameters</th></tr>
<tr><th colspan="2">Sustained</th><th>Free</th></tr>
<tr><td colspan="2">Maximum structural mass</td><td rowspan="9">Take-off configuration

Aircraft systems setting
(air conditioning)

v_1/v_r ratio

v_2/v_{SR} ratio</td></tr>
<tr><td colspan="2">Climb limited mass</td></tr>
<tr><td colspan="2">Obstacles</td></tr>
<tr><td>Runway</td><td>TODA
TORA
ASDA
Line-up corrections
Slope
Contaminants</td></tr>
<tr><td>Outside elements</td><td>Wind
Pressure
OAT</td></tr>
<tr><td colspan="2">Aircraft status</td></tr>
<tr><td colspan="2">Aircraft systems setting
(e.g. anti-icing)</td></tr>
<tr><td colspan="2">Brake energy capacity</td></tr>
<tr><td colspan="2">Tire speed limit</td></tr>
</table>

3.1 Aircraft configuration and systems setting

When it comes to the aircraft take-off configuration, it could be stated that while one configuration may positively contribute to performance on a short runway take-off, another leads to better climb gradients after the lift-off. The configuration giving the best performance for a particular take-off depends only on results of the optimization process. The influence of these parameters is more described in previous chapter.

The setting of some aircraft systems, mainly those which decrease the aircraft take-off performance through the reduction of the available engine power, may be unavoidable, because the operator can not influence the outside conditions, which require such setting. These are then considered to be sustained parameters. On the other hand, use of the air conditioning is only up to company policy, and unless the APU bleed is used, it is advised to switch it off during the take-off phase of a flight.

3.2 v_1/v_r ratio

3.2.1 v_1/v_r range

The definitions of speeds v_1 and v_r are stated in the beginning of chapter 2. According to these definitions, the decision speed v_1 must be logically less or equal than the rotation speed v_r. Therefore regulation sets the maximum value of this ratio to 1. The minimum of the v_1/v_r ratio is declared by the aircraft manufacturer, but it was proved by demonstrations that v_1 less than 84% of v_r presents no take-off performance advantages and that is why the range is most often identified as:

$$0{,}84 \leq v_1/v_r \leq 1$$

For a particular TOM, the value of v_r is fixed, as it directly depends on the aircraft mass. Therefore, any v_1/v_r increase (resp. decrease) should be considered to have the same effect on the take-off performance as a v_1 increase (resp. decrease) (7).

In addition to this, the regulation imposes requirements for both v_1 and v_r. According to CS-25.107 (a) the decision speed v_1 may not be less than v_{MCG} (minimum control speed on ground) to provide with a sufficient efficiency to control the aircraft in case of engine failure after reaching v_1. When it comes to the rotation speed v_r, the article (e) states that it may not be less than 105% of v_{MCA} (minimum control speed in the air) and also determined speed that allows reaching v_2 before a height of 35 ft above the take-off surface.

3.2.2 v_1/v_r ratio influence

The objective of this subchapter is to study the variation of the v_1/v_r ratio and its influence on the MTOM. The MTOM is limited by several parameters presented in previous as sustained and free. For the purpose of this chapter, it is necessary to assume all of them as fixed.

The influence of the v_1/v_r ratio change on different limitations of the MTOM is translated step-by-step by following figures. The first of them represents the runway limitations.

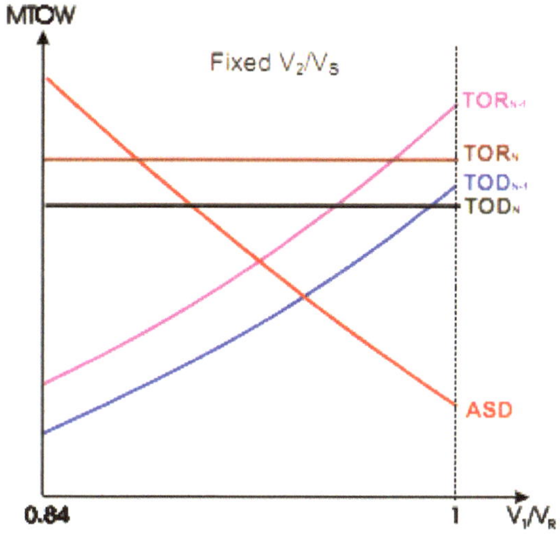

Figure 19: v_1/v_r effect on runway limited MTOM [Source: Airbus, Getting to grips of aircraft performance]

A higher v_1/v_r ratio (higher v_1) leads to a higher percentage of the acceleration phase with all engines operating during one engine failure take-off; consequently, it will take less time and distance to achieve v_2 at 35 ft (11). Less distance therefore supposes less restrictive limitations of the take-off run with one engine inoperative (TOR_{n-1}) and the take-off distance with one engine inoperative (TOD_{n-1}). The TOD and the TOR with all engines operating are independent on the v_1/v_r ratio, because all engines are working and there is no consequence on distance and time to reach 35 ft at speed v_2. On the contrary, due to prolonged distance during the acceleration phase to reach a higher v_1, a distance to allow braking after the rejected take-off is reduced. This results in more restrictive ASD limitation.

Figure 20 translates the influence of the v_1/v_r ratio on the climb, obstacle, brake energy and tire speed limited MTOM.

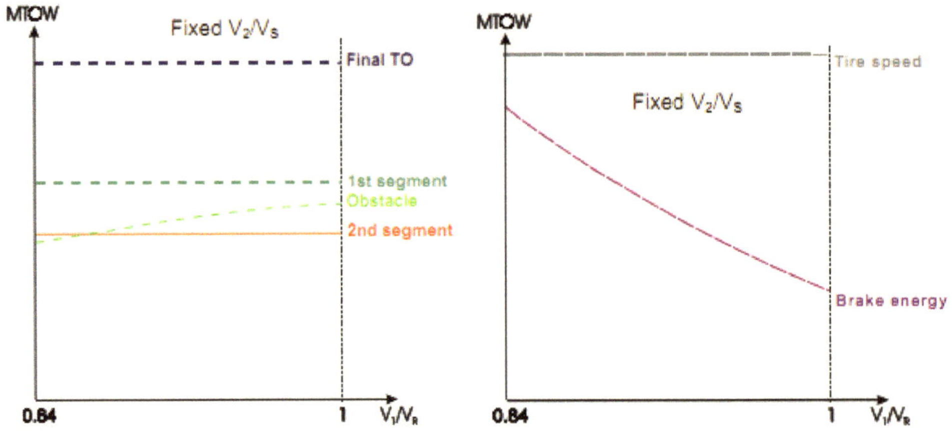

Figure 20: v_1/v_r effect on climb, obstacle, brake energy and tire speed limited MTOM [Source: Airbus, Getting to grips of aircraft performance]

The decision speed on ground v_1 has no direct influence on the flight climb gradients. This is the reason why neither first, second nor final take-off climb gradient limitation is influenced by the change of the v_1/v_r ratio. On the contrary, a higher v_1 supposes that in case of engine failure, the start of the take-off flight path is obtained at a shorter distance. This returns a prolonged flight distance to the obstacle and consequently lower gradient requirement to clear it.

As it could be observed form Figure 20, the tire speed limitation is independent from the v_1/v_r ratio change, but it is not true when it comes to the brake energy limitation. As this limitation depends on a kinetic energy of the aircraft, there is calculated maximum v_1 speed for each TOM. In general, the higher decision speed is the more restrictive limitation needs to be applied.

3.3 v_2/v_{SR} ratio

For a purpose of this subsection it is necessary to mention what v_{SR} is. The reference stall speed v_{SR} according to CS-25.103 is a calibrated airspeed, which should not be less than a 1-g stall speed. v_{SR} is expressed as:

$$v_{SR} \geq \frac{v_{CLMAX}}{\sqrt{n_{zw}}}$$

v_{CLMAX} is defined as a calibrated airspeed obtained when a lift coefficient is at maximum and n_{zw} is a load factor normal to the flight path at v_{CLMAX}.

3.3.1 v_2/v_{SR} range

The take-off climb speed v_2 is defined as a minimum calibrated airspeed that must be reached at a height of 35 ft above the runway surface, in case of critical engine failure. Moreover, CS-25.107 imposes conditions for determining the minimum value of v_2. In terms of calibrated airspeed it may not be less than:

- 1,13 v_{SR} for:
  - (i) Two-engined and three-engined turbo-propeller powered aeroplanes;
  - (ii) Turbojet powered aeroplanes without provisions for obtaining a significant reduction in the one-engine-inoperative power-on stall speed;
- 1,08 v_{SR} for:
  - (i) Turbo-propeller powered aeroplanes with more than three engines;
  - (ii) Turbojet powered aeroplanes with provisions for obtaining a significant reduction in the one-engine-inoperative power-on stall speed;
- 1,1 v_{MCA} (minimum control speed in the air) (8).

When it comes to a maximum value of v_2 (and consequently a maximum v_2/v_{SR} ratio), too high v_2 speed requires long take-off distances and leads to a reduction of the climb performance, which presents no advantage. Therefore, the maximum v_2/v_{SR} ratio is declared by the manufacturer and it is proved by demonstrations.

$$(v_2/v_{SR})_{min} \leq v_2/v_{SR} \leq (v_2/v_{SR})_{max}$$

Both the minimum and maximum value of the v_2/v_{SR} ratio depends on the aircraft type, but in general, any v_2/v_{SR} increase (resp. decrease) should be considered to have the same effect on the take-off performance as a v_2 increase (resp. decrease) (7).

3.3.2 v_2/v_{SR} ratio influence

This subchapter studies the influence of the v_2/v_{SR} ratio variations on the take-off performance, specifically the value of the MTOM. For the purpose of this, it is necessary to assume that all other sustained and free parameters are fixed.

The Figure 21 translates an increase of the v_2/v_{SR} ratio and its influence on the runway, tire speed and brake energy limitation of the MTOM.

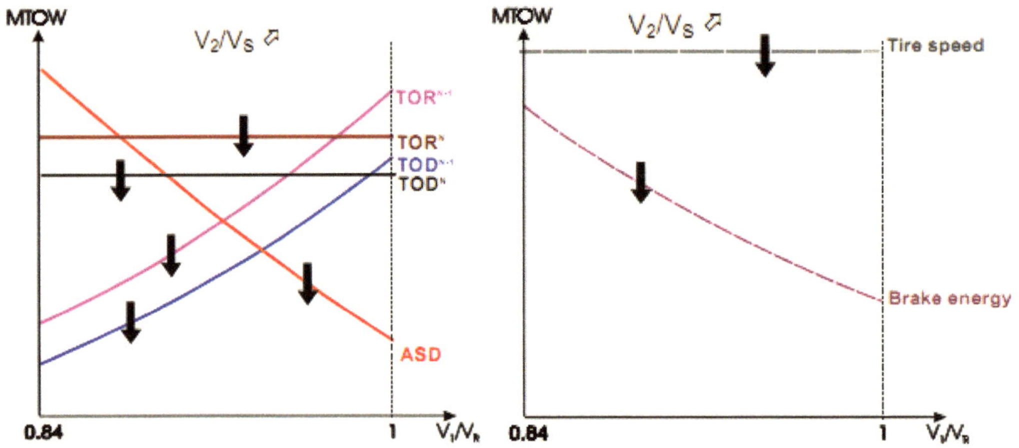

Figure 21: v_2/v_{SR} effect on runway, brake energy and tire speed limited MTOM [Source: Airbus, Getting to grips of aircraft performance]

For a given v_1/v_r ratio, any increase in the v_2/v_{SR} ratio leads to an increase of all take-off distances. This is a consequence of the prolonged acceleration phase, which is necessary to achieve a higher v_2 speed at 35 ft above the runway surface. Although v_2 speed has no direct impact on the ASD and brake energy limitation, the higher v_2 results in a higher v_r and therefore for given v_1/v_r ratio a higher v_1. This may lead to increased ASD requirement as well as more restrictive brake energy limitation. Similarly, with increased v_2 and consequently v_{LOF}, the tire speed limitation becomes more restraining. All of these, prolonged take-off distances, effects on the ASD, brake energy and tire speed limitation, thus negatively influence the value of the MTOM.

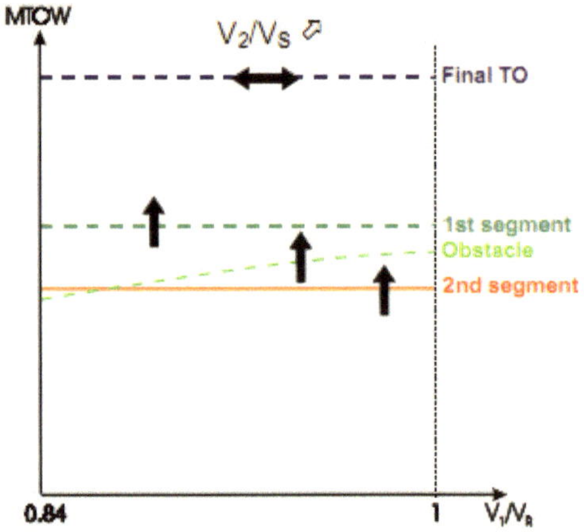

Figure 22: v_2/v_{SR} effect on climb, obstacle limited MTOM [Source: Airbus, Getting to grips of aircraft performance]

As it was discussed in previous subchapter, the aircraft manufacturer specifies the maximum value of the v_2/v_{SR} ratio. A flight at such v_2 speed results in best climb gradients during the 1st and 2nd take-off segment. Consequently, any v_2/v_{SR} ratio increase, up to its maximum value, results in a better climb limited MTOM for these two segments. On the contrary, the climb gradient at the final take-off segment is not influenced by v_2 variation as it is flown at final segment speed.

The improvement of the climb gradients supposes also improved obstacle limited MTOM. This is always true for distant obstacles, but may not be in case of relatively close obstacles. Due to necessity to prolong the take-off distance, the clearance from such obstacles may be reduced leading to the MTOM restrictions.

Table 5: Influence of v_2/v_{SR} ratio change on take-off limitations [Source: Aguiar, Aeroplane Performance]

(Assuming a fixed V_1/V_R)	When V_2/V_S Increases	
Limitation	Length/Clearance/Climb gradient/Speed/Energy	Weight limited by
ASDA	↗	↘
TORA/TODA	↗	↘
Close Obstacle	↘	↘
Distant Obstacle	↗	↗
1st Segment Gradient	↗	↗
2nd Segment Gradient	↗	↗
Final Segment Gradient	-	-
Tire Speed	↗	↘
Brake Energy	-	↘

3.4 Take-off data determination

3.4.1 MTOM

After considering all influences of the v_1/v_r ratio change (for fixed v_2/v_{SR} ratio), it is possible to determine an optimum MTOM and associated optimum v_1/v_r ratio for a specific take-off conditions. Following figures show that the optimum point always corresponds to the intersection of two limitation curves.

As described in previous, for specific aircraft take-off configuration both variations of the v_1/v_r and v_2/v_{SR} ratio result in the change of the MTOM limitations. Aim of the optimization process is to find an optimum value for both ratios. In other words, the MTOM and associated v-speeds are determined through an iterative process, which looks for the optimum v_1/v_r for a given v_2/v_{SR} and then for the optimum v_2/v_{SR} for that v_1/v_r (12). This process should continue until specific precision is achieved.

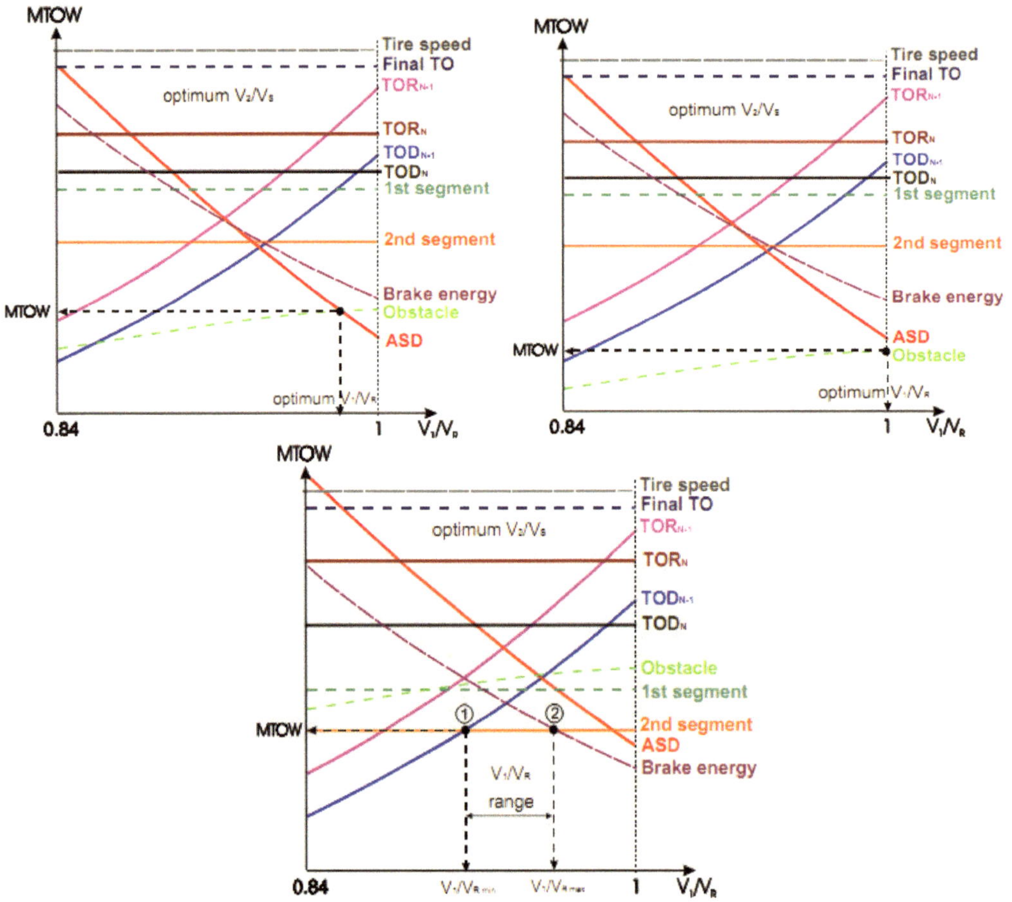

Figure 23: Optimum MTOM and v_1/v_r ratio [Source: Airbus, Getting to grips of aircraft performance]

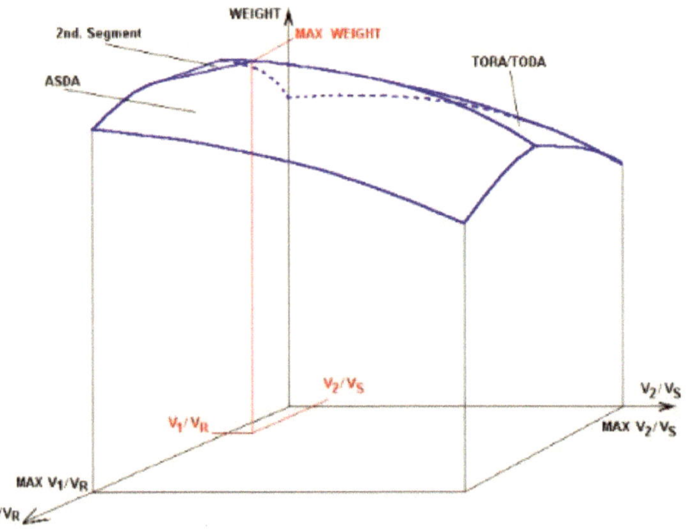

Figure 24: MTOM as function of v_1/v_r and v_2/v_{SR} [Source: Aguiar, Aeroplane Performance]

3.4.2 v-speeds

After obtaining the MTOM and appropriate v_1/v_r and v_2/v_{SR} ratios, it is possible to calculate all required v-speeds according to the AFM data. As it could be observed these speeds are never fixed values because both the aircraft stall and the rotation speed depend on the aircraft weight.

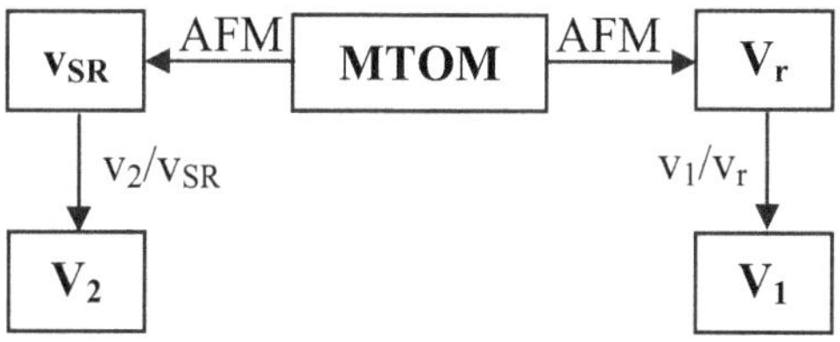

Figure 25: Take-off speed calculation

4 EXISTING RUNWAY ANALYSES PRODUCTS

Prior each take-off it is necessary to consider all influencing factors described in previous chapters. Computing the optimized MTOM and appropriate v-speeds only according to the AFM data and graphs before every single take-off would be extremely time-consuming. That is the reason why several companies developed specialized software to perform such calculation for various take-off conditions. This shortens the time required for the flight preparation. The output of this software is usually a table in either paper or electronic form.

The runway analyses, or sometimes called aerodrome analyses, are therefore the output form of the specific software. The intention of this chapter was to review existing runway analyses software products. As these are always paid, the airline-software providing companies or airline companies themselves were not always willing to give any more specific information. The search was conducted through a personal contact, e-mail correspondence or web search, many times with a great help of specialized websites, for instance www.airlinesoftware.net. The selection of following products was based on a willingness of the companies to provide with exemplary analyses, their prestige and airlines using a particular software solution. The objective of this chapter is then to make a review of selected products and compare them.

Navigation through the runway analyses tables should be as easiest-to-follow as possible, but sometimes it could be a little bit confusing. Therefore, another objective of this chapter is to describe an appearance of the runway analyses tables and to find out the most appropriate layout to find a proper value of the MTOM and v-speeds for a particular runway and conditions.

4.1 Aircraft manufacturer software

All aircraft manufactures are required to provide aircraft operators with the AFM including the up-to-date performance and operational characteristics. Based on these characteristics, mostly in a form of graphs, the operator should be able to ensure a safe operation. As mentioned previously, it is a long and tedious way and therefore many aircraft manufactures offer their take-off performance software themselves.

Example of such manufacturer is Airbus SAS. Its TLO module (Take-off and Landing Optimization) runs under PEP (Performance Engineering Programs). TLO is an interface common to all aircraft types, but on the other hand the calculation program depends on a specific aircraft type. There are used two types of programs - TLC (Take-off and Landing Computation) and OCTOPUS (Operational and Certified Take-off and landing Universal Software). Both of these provide with so called RTOW charts (Regulatory Take-off Weight) in different formats.

The older of these two programs TLC was developed in 1980's and is used for the A300, A310 and some models of A320 calculation. The header of the output table states a take-off configuration of the aircraft, departure aerodrome with its ICAO code and a runway to be used; these suppose the table to be used. Right under it, there is specified aerodrome information as elevation, length of the TORA, ASDA, TODA and a runway slope. Next to it, there are specified codes of limitations which could affect a particular take-off. Along with these, the right upper part of the TLC output table involves the aircraft type, its setting (for example setting of the air conditioning, anti-icing system or use of reverse) and a runway condition. When it comes to main output data, the MTOM and v-speeds, these are organized in cells, each related to a particular temperature and wind. According to this right to left and up to down order it is easy to find the appropriate cell. The first line of this cell represents the MTOM followed with a code of limitations affecting the take-off. The MTOM could be presented in both kilograms or pounds and it is important to pay attention which unit is used. The second line gives values of the v-speeds in order v_1, v_r and v_2. These values are usually presented in knots as IAS. Minimum and maximum acceleration heights, respectively their adequate values expressed in QNH altitude, are located at the footer of the TLC table.

The newer program OCTOPUS, unlike the TLC, is no longer based on pre-computed data, but enables the performance computation based on physics equations of the airplane and engine characteristics. This benefits in a more accurate MTOM. Similar to the TLC table, header specifies the departure aerodrome, runway to be used, runway characteristics and setting of the aircraft systems. Moreover, the number of relevant obstacles is stated under the runway design number. Location of the MTOM and appropriate v-speeds for a specific temperature and wind is based on a similar table with cells. Main advantage is a presence of the corrections for several runway conditions, delta pressure and the aircraft system settings. The bottom part of this analysis includes also the values of the minimum and maximum acceleration heights and altitudes. The program OCTOPUS is used for performance calculation for the A318, A319, some types of A320, A321, A330, A340 and A380.

Nowadays, based on the concept of the Less Paper Cockpit (LPC), the traditional paper charts are being replaced by using of laptops in the cockpit. This additionally reduces the flight preparation time and the risk of error. This also changes the appearance of the calculation program output. Nevertheless, the computation could be still based on the same software. It is important to mention that although there are several companies producing runway analyses, many aircraft operators have their own tailored software. It is usually based on the calculation program of the aircraft manufacturer and only user interface is adjusted for company needs. Example of such software and company is TAP Portugal and their Take-off and Landing Performance Application, which is nowadays being implemented into ground operation and runs under the Airbus' OCTOPUS.

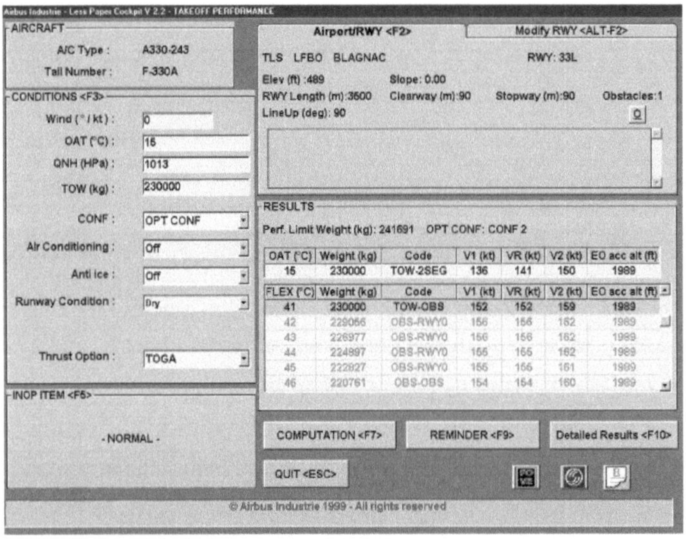

OAT (°C)	Weight (kg)	Code	V1 (kt)	VR (kt)	V2 (kt)	EO acc alt (ft)
15	230000	TOW-2SEG	136	141	150	1989

FLEX (°C)	Weight (kg)	Code	V1 (kt)	VR (kt)	V2 (kt)	EO acc alt (ft)
41	230000	TOW-OBS	152	152	159	1989
42	229066	OBS-RWY0	156	156	162	1989
43	226977	OBS-RWY0	156	156	162	1989
44	224897	OBS-RWY0	155	155	162	1989
45	222827	OBS-RWY0	155	155	161	1989
46	220761	OBS-OBS	154	154	160	1989

Figure 26: Airbus LPC take-off interface

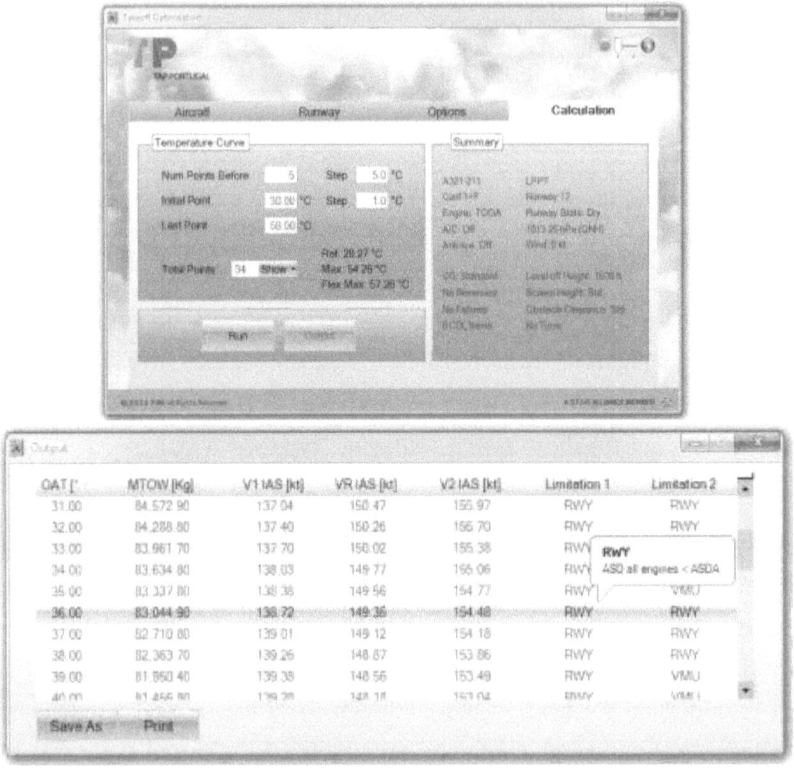

OAT [°	MTOW [Kg]	V1 IAS [kt]	VR IAS [kt]	V2 IAS [kt]	Limitation 1	Limitation 2
31.00	84.572.90	137.04	150.47	155.97	RWY	RWY
32.00	84.288.80	137.40	150.26	156.70	RWY	RWY
33.00	83.961.70	137.70	150.02	155.38	RWY	[illegible]
34.00	83.634.80	138.03	149.77	155.06	RWY	[illegible]
35.00	83.337.80	138.38	149.56	154.77	RWY	VMU
36.00	83.044.90	138.72	149.36	154.48	RWY	RWY
37.00	82.710.80	139.01	149.12	154.18	RWY	RWY
38.00	82.363.70	139.26	148.87	153.86	RWY	RWY
39.00	81.950.40	139.38	148.56	153.49	RWY	VMU
40.00	81.456.80	139.28	148.18	153.04	RWY	VMU

Figure 27: TAP Portugal TLP take-off optimization mode interface

CONF1+F — TOULOUSE - BLAGNAC — LFBO 33L

ELEVATION= 497.0(FT) T.O.R.A. = 3500.0(M) A.S.D.A. = 3560.0(M) T.O.D.A. = 3500.0(M) SLOPE = 0.00(%)	CODES LIMITATIONS 1-STRUCTURE 5-TYRE SPEED 2-2ND-SEGMENT 6-BRAKE ENERGY 3-RUNWAY 7-RWY 2 ENGINES 4-OBSTACLE 8-FINAL T.O.	A320-214/AA/565B4IMP JAR MC69 15-Jan-02 QNH=1013.25 (MB)
		AIR COND. OFF CG25% ANTI-ICING OFF
TREF = 43 DC TMAX = 54 DC	WIND (KT) MTOW (KG) CODES MEAN V1 - VR - V2 (IAS.KT)	WITHOUT REVERSE DRY RUNWAY

O A T DC	-10	-5	0	10	20
-31.0	83130 2-6 143 159 161	84344 2-6 148 164 166	85000 2-1 152 167 168	85000 2-1 153 167 168	85000 2-1 154 167 168
-17.0	82032 2-6 140 156 158	83303 2-6 145 161 162	84509 2-6 151 165 167	85000 2-1 154 167 169	85000 2-1 155 167 169
-2.0	80554 6-6 137 154 156	82198 2-6 142 157 159	83450 2-6 148 162 164	84213 2-6 151 165 166	84931 2-6 155 168 169
13.0	78844 6-6 135 152 154	80985 6-6 139 155 156	82358 2-6 144 159 160	83186 2-6 148 162 163	83929 2-6 151 165 166
28.0	77179 6-6 133 150 152	79329 6-6 137 153 155	81274 2-6 142 156 157	82142 2-6 145 159 160	82920 2-6 148 161 163
43.0	75603 6-6 131 149 151	77764 6-6 135 151 153	79921 6-6 139 154 155	81094 2-6 142 156 157	81922 2-6 146 159 160
45.0	74924 6-6 132 148 150	77109 6-6 136 151 152	79170 2-6 141 153 155	80022 2-6 144 156 158	80378 2-8 146 158 159
46.0	74602 6-6 132 148 150	76798 6-6 137 150 152	78656 2-6 141 154 155	79483 2-6 145 157 158	79543 2-8 146 157 158
48.0	73985 6-6 133 147 149	76207 6-6 137 150 151	77557 2-6 143 154 156	77872 8-6 147 156 157	77872 2-8 146 156 157
49.0	73690 6-6 133 147 149	75752 2-6 138 150 152	77013 2-6 144 155 156	77036 2-8 145 155 156	77036 2-8 146 155 156
51.0	73137 6-6 134 147 148	74692 2-6 140 151 152	75364 2-8 143 153 155	75364 2-8 145 153 155	75364 2-8 144 153 155
53.0	72340 2-6 135 147 148	73600 2-6 141 152 153	73691 2-8 143 152 153	73691 2-8 143 152 153	73691 2-8 142 152 153
54.0	71830 2-6 136 147 149	72854 2-8 141 151 153	72854 2-8 143 151 153	72854 2-8 142 151 153	72854 2-8 140 151 153
56.0	70731 2-6 137 147 149	71888 2-6 143 152 153	72840 2-8 149 156 157	72840 2-8 150 156 157	72840 2-8 150 156 157
58.0	69645 2-6 139 147 149	70773 2-6 145 152 153	71795 2-6 151 157 158	72436 2-6 155 160 161	72824 2-2 158 162 163
59.0	69116 2-6 140 147 149	70209 2-6 146 152 153	71229 2-6 152 157 158	71837 2-6 156 160 161	72025 2-2 158 161 162
61.0	68037 2-6 141 148 149	69091 2-6 147 153 154	70040 2-6 153 157 158	70431 2-2 157 160 161	70431 2-2 157 160 161
62.0	67523 2-6 142 148 149	68529 2-6 148 153 154	69459 2-6 154 158 159	69635 2-2 156 159 160	69635 2-2 155 159 160
64.0	66443 2-6 143 149 150	67414 2-6 150 153 154	68040 2-2 155 157 158	68040 2-2 154 157 158	68040 2-2 153 157 158
66.0	65363 2-6 145 149 150	66271 2-6 151 154 155	66447 2-2 153 155 156	66447 2-2 152 155 156	66447 2-2 151 155 156
67.0	64817 2-6 146 149 150	65650 2-2 152 154 155	65650 2-2 151 154 155	65650 2-2 151 154 155	65650 2-2 150 154 155

/MINI. ACCELERATION HEIGHT : 516.(FT) QNH ALT. : 1013.(FT)
/MAXI. ACCELERATION HEIGHT : 2215.(FT) QNH ALT. : 2712.(FT)

Figure 28: TLC take-off chart example

A319114 - JAA	CFM56-5A5 engines	TOULOUSE - BLAGNAC	33L	19.0.0 15-JAN-02 AD114A03 *V10
QNH 1013.25 HPA Air cond. Off Anti-icing Off All reversers inoperative Dry check		Elevation 497 FT TORA 3500 M Isa temp 14 C TODA 3500 M rwy slope 0.00% ASDA 3560 M	0 obstacle	DRY CONF 1+F

OAT C	TAILWIND -10 KT	TAILWIND -5 KT	WIND 0 KT	HEADWIND 10 KT	HEADWIND 20 KT
0	78.3 3/6 151/57/62	79.6 3/6 157/62/67	80.0 3/7 158/64/69	80.0 3/7 156/64/69	80.0 3/7 154/64/69
10	77.7 3/6 148/55/60	78.9 3/6 155/60/65	80.0 6/7 161/65/69	80.0 3/7 159/65/69	80.0 3/7 157/65/69
20	76.9 3/6 146/53/58	78.2 3/6 152/58/63	79.4 3/6 159/63/68	80.0 6/7 163/66/71	80.0 3/7 161/66/71
30	76.2 3/6 144/51/56	77.5 3/6 150/56/61	78.7 3/6 156/61/66	79.4 3/6 160/64/69	80.0 6/7 164/67/72
32	76.1 3/6 144/51/56	77.4 3/6 150/56/61	78.6 3/6 156/60/65	79.3 3/6 160/63/68	80.0 3/6 164/67/71
34	76.0 3/6 143/51/56	77.3 3/6 149/55/60	78.5 3/6 155/60/65	79.2 3/6 159/63/68	79.9 3/6 164/66/71
36	75.9 3/6 143/50/55	77.2 3/6 149/55/60	78.4 3/6 155/59/64	79.1 3/6 159/63/68	79.7 3/6 163/66/71
38	75.1 3/6 143/50/55	76.3 3/6 149/55/60	77.5 3/6 155/59/64	78.2 3/6 159/62/67	78.8 3/6 164/66/70
40	74.2 3/6 144/50/55	75.4 3/6 150/55/60	76.5 3/6 156/59/64	77.2 3/6 160/62/67	77.8 3/6 164/66/70
42	73.3 3/6 145/50/55	74.5 3/6 151/55/59	75.5 3/6 157/59/64	76.2 3/6 161/62/67	76.7 3/6 165/66/70
44	72.4 3/6 146/50/55	73.6 3/6 152/55/59	74.6 3/6 157/59/64	75.2 3/6 162/62/67	75.6 3/6 165/65/70
46	71.5 3/6 146/50/55	72.6 3/6 152/54/59	73.5 3/6 158/59/64	74.1 3/6 162/62/67	74.4 2/3 164/64/69
48	70.6 3/6 147/50/55	71.6 3/6 153/55/59	72.5 3/6 159/59/64	72.8 2/3 161/61/66	73.1 2/3 163/63/67
50	69.6 3/6 148/50/55	70.6 3/6 154/55/59	71.3 2/3 158/58/63	71.5 2/3 160/60/65	71.8 2/3 162/62/66
	INFLUENCE OF RUNWAY CONDITION				
WET	-1.0 -3 -13/ -2/ -2 (+50) -1.0 -3 -13/ 0/ 0	-0.7 -2 -13/ -2/ -2 (+50) -1.3 -3 -13/ 0/ 0	-0.5 -2 -12/ -2/ -2 (+50) -1.1 -3 -12/ 0/ 0	-0.3 -1 -10/ -1/ -1 (+50) -0.3 -1 -10/ 0/ 0	-0.2 -1 -10/ -1/ -1 (+50) -0.2 -1 -10/ 0/ 0
D QNH HPA	**INFLUENCE OF DELTA PRESSURE**				
-10.0	-0.5 -2 0/ -1/ -1 (+50) -0.5 -2 0/ 0/ 0	-0.5 -2 0/ -1/ -1 (+50) -0.5 -2 0/ 0/ 0	-0.6 -2 0/ 0/ 0 (+50) -0.6 -2 0/ 0/ 0	-0.5 -1 0/ 0/ -1 (+50) -0.5 -1 0/ 0/ 0	-0.5 -1 0/ 0/ -1 (+50) -0.5 -1 0/ 0/ 0
+10.0	+0.2 0 +1/ 0/ 0 (+50) +0.2 0 +1/ 0/ 0	+0.2 0 +1/ 0/ 0 (+50) +0.2 0 +1/ 0/ 0	0.0 0 +1/ +1/ +1 (+50) 0.0 0 +1/ +1/ +1	0.0 0 +1/ +1/ +1 (+50) 0.0 0 +1/ +1/ +1	0.0 0 0/ 0/ 0 (+50) 0.0 0 0/ 0/ 0
	INFLUENCE OF AIR COND.				
On	-1.2 -3 0/ 0/ 0 (+50) -1.2 -3 0/ 0/ 0	-1.4 -3 0/ 0/ 0 (+50) -1.4 -3 0/ 0/ 0	-1.5 -3 0/ 0/ 0 (+50) -1.5 -3 0/ 0/ 0	-1.5 -3 0/ 0/ 0 (+50) -1.5 -3 0/ 0/ 0	-1.5 -3 0/ 0/ -1 (+50) -1.5 -3 0/ 0/ 0

LABEL FOR INFLUENCE DW (1000 KG) DTFLEX DV1-DVR-DV2 (KT) (TVMC LIMIT C) DW (1000 KG) DTFLEX DV1-DVR-DV2 (KT)	MTOW(1000 KG) codes V1min/VR/V2 (kt)	VMC LIMITATION	Tref (OAT) = 36 C Tmax (OAT) = 54 C	Min acc height 421 FT Max acc height 1950 FT	Min QNH alt 918 FT Max QNH alt 2447 FT
	LIMITATION CODES: 1=1st segment 2=2nd segment 3=runway length 4=obstacles 5=tire speed 6=brake energy 7=max weight 8=final take-off 9=VMU			Min V1/VR/V2 = 109/14/21 CHECK VMU LIMITATION Correct. V1/VR/V2 = 1.0 KT/1000 KG	

Figure 29: OCTOPUS take-off chart example

4.2 APG

Aircraft Performance Group, Inc. (APG) is American company providing several performance calculations for the aircraft operation such as weight and balance or the take-off and landing data. Their products are based on internet applications or stand-alone systems and therefore made available by using several electronic devices.

For the runway analyses calculation the APG uses a program called ATLAS. This software is available via the internet and the output form is adjusted according to the platform used. Typical way to obtain the take-off performance data is by using a laptop. After sending a request for a particular aerodrome runway analysis, these data are computed and presented in form of a table, which could be printed as a classic paper runway analysis. To comparison with previously discussed tables, this one has not as much easy-to-follow format. Some information (for example airport name, the aircraft type and its configuration, the runway length corresponding to the TORA and slope) are easy to find, but a problem may occur when it comes to determination of the MTOM. This format offers data for two runways at one page and therefore it is faster to compare them and decide for more convenient departure runway. On the other hand, neither of these values could be correct. The particular mass for a specific temperature is always necessary to compare with a climb limited mass in the rightmost column. Therefore, the correct value of the MTOM is always the lower of these figures. Moreover, all of the stated figures are considered for a zero wind. If a nonzero wind component in a direction of the runway is present, the appropriate corrections stated under the table must be applied. Similarly, the corrections for various runway contaminants and aircraft statuses are situated under the table. The last line specifies the acceleration altitude for a particular departure. The APG provides their analyses for either normal or icing conditions, because in such cases the AFM requires a different ratio between v_2 and v_{SR}. If special one-engine flight procedures need to be applied during the departure from a particular runway, these are stated on separate paper.

As mentioned previously, the APG provides their product also through the other electronic devices such as smart phones or tablets. The output data are presented in much simpler way decreasing the possibility of errors and provide with additional information, for instance weather and NOTAMs. The APG computing software powers also some websites offering pilots help with their pre-flight preparation; an example of such website is www.fltplan.com.

-- BTS -LZIB -- TAKEOFF PERFORMANCE -- BTS -LZIB --
ATR-72 BRATISLAVA, SVK
ELEVATION 436 PW-124B ENG M.R. STEFANIK
V2 = 1.20 VS
TAKEOFF FLAPS 15.0 DEGREES VCLIMB = 1.18 VS

*** NORMAL CONDITIONS ***

RUNWAY		04	22	
LENGTH		2899	2899	CLIMB
SLOPE		-0.06	0.06	LIMIT
TMP C	TORQ %	LIMIT WEIGHT(KGS)/LIMIT CODE V1/VR/V2(KIAS)		
-20	90.0	21999/ST 117/117/119	21999/ST 117/117/119	22000.
-15	90.0	21999/ST 117/117/119	21999/ST 117/117/119	22000.
-10	90.0	21999/ST 117/117/119	21999/ST 117/117/119	22000.
-5	90.0	21999/ST 117/117/119	21999/ST 117/117/119	22000.
0	90.0	21999/ST 117/117/119	21999/ST 117/117/119	22000.
5	90.0	21999/ST 117/117/119	21999/ST 117/117/119	22000.
10	90.0	21999/ST 117/117/119	21999/ST 117/117/119	22000.
12	90.0	21999/ST 117/117/119	21999/ST 117/117/119	22000.
14	90.0	21999/ST 117/117/119	21999/ST 117/117/119	22000.
16	90.0	21999/ST 117/117/119	21999/ST 117/117/119	21999.
18	90.0	21999/ST 117/117/119	21999/ST 117/117/119	21941.
20	90.0	21999/ST 117/117/119	21999/ST 117/117/119	21883.
22	90.0	21999/ST 117/117/119	21999/ST 117/117/119	21825.
24	90.0	21999/ST 117/117/119	21999/ST 117/117/119	21766.
26	90.0	21999/ST 117/117/119	21999/ST 117/117/119	21708.
28	90.0	21999/ST 117/117/119	21999/ST 117/117/119	21650.
30	90.0	21943/-O 117/117/119	21943/-O 117/117/119	21591.
32	90.0	21830/-O 117/117/119	21830/-O 117/117/119	21513.
34	89.3	21621/-O 117/117/119	21621/-O 117/117/119	21319.
36	88.5	21246/-O 117/117/119	21246/-O 117/117/119	20968.
38	86.8	20885/-O 116/116/118	20885/-O 116/116/118	20621.
40	85.0	20489/-O 115/115/117	20489/-O 115/115/117	20289.
45	80.6	19619/-O 112/112/115	19619/-O 113/113/115	19506.
47	78.8	19249/-O 111/111/114	19250/-O 111/111/114	19179.

HW +KGS/KT	0	0	
TW -KGS/KT	254	253	
1 EEC OFF-KGS	0	100	4450
ATPCS OFF-KGS	370	520	3750
ATO FTHR INP-KGS	40	210	3800
1 ACWGEN INP-KGS	0	0	2750
GEAR DOWN -KGS	0	0	2750
ANTI SKD INP-KGS	0	0	
ONE BRK INP-KGS	2670	2660	
WATER 1/8 " -KGS	0	0	
SLUSH 1/4 " -KGS	0	0	
SLUSH 1/2 " -KGS	0	0	
COMPCT SNOW -KGS	0	0	
ICY RUNWAY -KGS	0	0	
TYPE II/IV -KGS	0	0	
ACCEL HT (MSL)	1230	1230	

*** OBSERVE STRUCTURAL LIMITS *** 25Feb10

Figure 30: APG take-off chart example

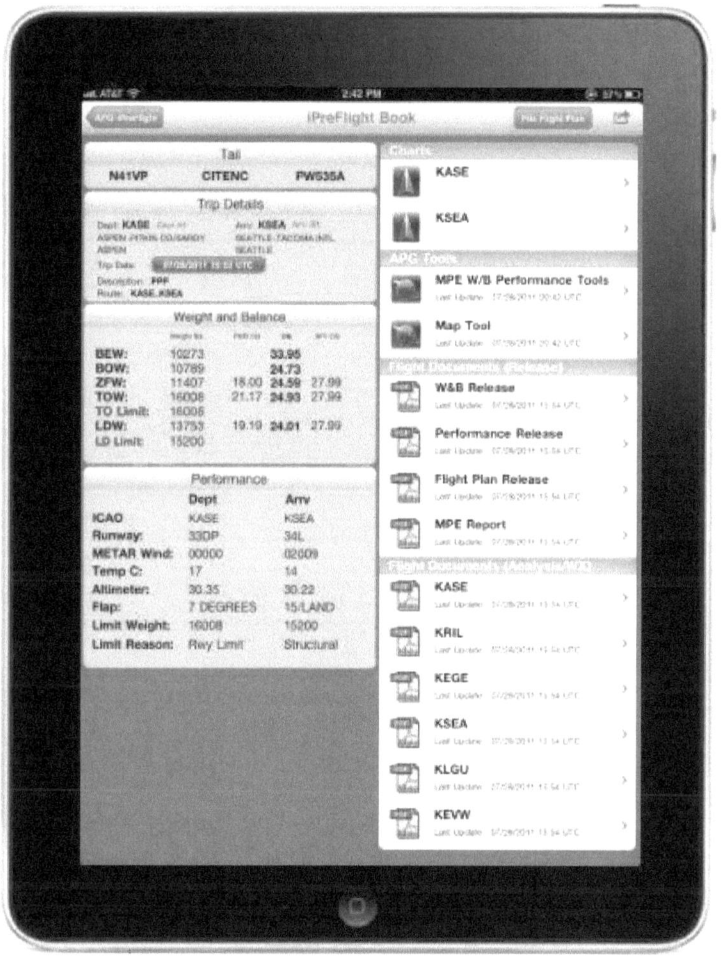

Figure 31: APG performance data tablet format example [Source: www.apg.aero; online: 25.2.2012]

4.3 Flygprestanda

Flygprestanda AB is a Swedish company specializing in the support of the take-off and landing calculations. Their runway analyses products are computed by themselves developed software called Guru. Based on a concept of the LPC, the Flygprestanda nowadays provide their products also in an electronic form for several platforms, but still also produce runway analyses in a traditional paper format.

The header in the paper format specifies the aircraft type, configuration and the airport characteristics, but rest of the runway analyses is organized in several tables. These tables are arranged in temperature to wind layout and each one of them represents different state of the runway. Therefore, no additional corrections need to be considered as the user is able to find out the exact value of the MTOM and appropriate v-speeds for a dry, wet or reduced friction characteristics runway. When it comes to a contaminated runway, water equivalent depth conversion table has to be used to obtain the correct values. This approach also benefits in reducing the possibility of errors and making the Flygprestanda runway analyses one of the easiest-to-follow. In addition to those data, information about all relevant obstacles included in the calculation are presented (the distance from the brake release point and the height above the runway end) advising pilots about a potential hazard.

Flygprestanda AB Performance Engineering www.flygp.se	ATR72-500 / PW127-F Route Performance Manual BOOST OFF	Section 3 Sheet LZIB-04 Date Nov 11 2011
TORA 2900 m ASDA 2900 m TODA 3200 m LDA 2900 m	BRATISLAVA/M.R. Stefanik Slope -0.07% Downhill AD Elev 436 ft LZIB/BTS 04	Normal cond. Aircond OFF Flaps 15

No note!

DGAC appr AFM used **Optimum V1/Vr** **Optimum V2/Vs**

Anti-ice OFF **Takeoff - Dry runway -** (All masses in kg) **Check MTOM**

OAT	T 10	T 5	0 (Calm)	H 5	H 10	H 15	H 20
-10°C	23000(28) [109 109 114]	23000(0) [109 109 114]	23000(0) [109 109 114]	23000(0) [109 109 114]	23000(0) [109 109 114]	23000(0) [109 109 114]	23000(0) [109 109 114]
-5°C	23000(28) [108 109 114]	23000(0) [109 109 114]	23000(0) [109 109 114]	23000(0) [109 109 114]	23000(0) [109 109 114]	23000(0) [109 109 114]	23000(0) [109 109 114]
0°C	23000(28) [106 110 114]	23000(0) [110 110 114]	23000(0) [110 110 114]	23000(0) [110 110 114]	23000(0) [110 110 114]	23000(0) [110 110 114]	23000(0) [110 110 114]
5°C	23000(29) [105 110 114]	23000(0) [110 110 114]	23000(0) [110 110 114]	23000(0) [110 110 114]	23000(0) [110 110 114]	23000(0) [110 110 114]	23000(0) [110 110 114]
10°C	23000(29) [103 110 114]	23000(0) [110 110 114]	23000(0) [110 110 114]	23000(0) [110 110 114]	23000(0) [110 110 114]	23000(0) [110 110 114]	23000(0) [110 110 114]
15°C	23000(29) [101 110 114]	23000(16) [110 110 114]	23000(0) [110 110 114]	23000(0) [110 110 114]	23000(0) [110 110 114]	23000(0) [110 110 114]	23000(0) [110 110 114]
20°C	23000(29) [100 110 114]	23000(24) [110 110 114]	23000(0) [110 110 114]	23000(0) [110 110 114]	23000(0) [110 110 114]	23000(0) [110 110 114]	23000(0) [110 110 114]
25°C	23000(29) [98 110 114]	23000(29) [110 110 114]	23000(2) [110 110 114]	23000(2) [110 110 114]	23000(2) [110 110 114]	23000(2) [110 110 114]	23000(2) [110 110 114]
30°C	23000(28) [97 110 114]	23000(29) [108 110 114]	23000(15) [110 110 114]	23000(15) [110 110 114]	23000(15) [110 110 114]	23000(15) [110 110 114]	23000(15) [110 110 114]
35°C	22936(27) C [96 113 116]	23000(29) [107 113 117]	23000(24) [113 113 117]	23000(24) [113 113 117]	23000(24) [113 113 117]	23000(24) [113 113 117]	23000(24) [113 113 117]
40°C	22149(24) C 96 113 116	22685(25) C [106 124 127]	22736(25) C [116 128 131]	22736(25) C [120 128 131]	22736(25) C [124 128 131]	22736(25) C [127 128 131]	22736(25) C [128 128 131]
45°C	21513(23) F 97 115 118	21826(24) C 107 126 129	21827(24) C 114 126 129	21827(24) C 117 126 129	21827(24) C 121 126 129	21827(24) C 124 126 129	21827(24) C 126 126 129

Anti-ice OFF **Takeoff - Wet runway -** (All masses in kg) **Rev. OPER** **Check MTOM**

OAT	T 10	T 5	0 (Calm)	H 5	H 10	H 15	H 20
-10°C	23000(28) [109 109 114]	23000(0) [109 109 114]	23000(0) [109 109 114]	23000(0) [109 109 114]	23000(0) [109 109 114]	23000(0) [109 109 114]	23000(0) [109 109 114]
-5°C	23000(28) [108 109 114]	23000(0) [109 109 114]	23000(0) [109 109 114]	23000(0) [109 109 114]	23000(0) [109 109 114]	23000(0) [109 109 114]	23000(0) [109 109 114]
0°C	23000(28) [106 110 114]	23000(0) [110 110 114]	23000(0) [110 110 114]	23000(0) [110 110 114]	23000(0) [110 110 114]	23000(0) [110 110 114]	23000(0) [110 110 114]
5°C	23000(29) [105 110 114]	23000(0) [110 110 114]	23000(0) [110 110 114]	23000(0) [110 110 114]	23000(0) [110 110 114]	23000(0) [110 110 114]	23000(0) [110 110 114]
10°C	23000(29) [103 110 114]	23000(0) [110 110 114]	23000(0) [110 110 114]	23000(0) [110 110 114]	23000(0) [110 110 114]	23000(0) [110 110 114]	23000(0) [110 110 114]
15°C	23000(29) [101 110 114]	23000(16) [110 110 114]	23000(0) [110 110 114]	23000(0) [110 110 114]	23000(0) [110 110 114]	23000(0) [110 110 114]	23000(0) [110 110 114]
20°C	23000(29) [100 110 114]	23000(24) [110 110 114]	23000(0) [110 110 114]	23000(0) [110 110 114]	23000(0) [110 110 114]	23000(0) [110 110 114]	23000(0) [110 110 114]
25°C	23000(29) [98 110 114]	23000(29) [110 110 114]	23000(2) [110 110 114]	23000(2) [110 110 114]	23000(2) [110 110 114]	23000(2) [110 110 114]	23000(2) [110 110 114]
30°C	23000(28) [97 110 114]	23000(29) [108 110 114]	23000(15) [110 110 114]	23000(15) [110 110 114]	23000(15) [110 110 114]	23000(15) [110 110 114]	23000(15) [110 110 114]
35°C	22936(29) C [96 113 116]	23000(29) [107 113 117]	23000(24) [113 113 117]	23000(24) [113 113 117]	23000(24) [113 113 117]	23000(24) [113 113 117]	23000(24) [113 113 117]
40°C	22149(30) C 95 113 116	22685(33) C [106 124 127]	22736(25) C [116 128 131]	22736(25) C [120 128 131]	22736(25) C [124 128 131]	22736(25) C [127 128 131]	22736(25) C [128 128 131]
45°C	21513(28) F 96 115 118	21826(24) C 103 126 129	21827(24) C 114 126 129	21827(24) C [illegible] 126 129	21827(24) C 121 126 129	21827(24) C 124 126 129	21827(24) C 126 126 129

Limit code: F=Field, O=Obstacle, Oz=10 min thrust limit, B=Brakes, and C=Climb.
OPS 1 line-up with 90 degr entry

Obstacles included in calculation: (Height above runway end / Distance from brake release point)
Oa: 43 ft/3916 m Ob: 226 ft/15500 m Oc: 92 ft/9701 m

Approach climb limited mass. 2.5% (All masses in kg) **Check MLM**

Flaps	-10°C	-5°C	0°C	5°C	10°C	15°C	20°C	25°C	30°C	35°C	40°C	45°C
15°	22500	22500	22500	22500	22500	22500	22500	22500	22500	22385	21551	20749

Landing (Dry / Wet surface.) Stop within 70% of LDA.(All masses in kg) **Check MLM**

Flaps	T 10	T 5	0 (Calm)	H 5	H 10	H 15	H 20
30°	22500 / 22500	22500 / 22500	22500 / 22500	22500 / 22500	22500 / 22500	22500 / 22500	22500 / 22500

[atr72500 v3.15 by Flygprestanda AB] obstmet 2 [AT72-3] rev 02

Figure 32: Flygprestanda take-off chart example

4.4 ASAP

The abbreviation ASAP stands for Automated Systems in Aircraft Performance, Inc., which is an American company situated in state Pennsylvania. This company specializes in providing the aircraft take-off and landing information, based on maximum masses and speeds, with data driven by specific runway characteristics and condition. Their software called Star enables to offer their customers the performance data in both electronic and paper form.

The ASAP paper form runway analyses inform at their heading about the aircraft type, its configuration, general airport characteristics and conditions, which the particular analysis was computed for. The advantage of these analyses is an overview of up to six different runways at one page and therefore faster way to find out the most convenient runway for particular take-off conditions by a comparison of the MTOMs. One of the drawbacks, similar to APG analyses, is a fact that the value of the MTOM stated for particular temperature and runway may not be the correct one. This figure has to be always compared with a value in the last column, which informs about a climb limited mass and therefore the lower one of these must be taken into consideration. To avoid possible mistakes, in the ASAP analyses an asterisk mark appears next to the runway limited mass to indicate that it is higher than the climb limited mass. Information about the runway length and slope are situated along with wind and pressure corrections and the acceleration altitudes right under the table. The fact that the corrections are not directly involved into the table and require another calculation may lead to possible errors. At the very bottom of the ASAP analyses special non-standard departure procedures for engine failure are described. Other significant notes may appear here as well. When it comes to a wet or contaminated runway, the corrections for such conditions are not present and a special table is produced for each of such circumstances.

NATIONAL AIRLINES RPM REV DATE: 01-Jul-2011 — **LAS VEGAS, NV. MCCARRAN INTL**

– KLAS –	FLAPS 10	CF6-80C2B1F
ELEV. 2181 FT	B747-400	– LAS –
MAX TEMP 50 C	TAKE OFF PERFORMANCE - KGS	MCCARRAN INTL
CF6-80C2B1F ENGINE	DRY THRUST - A/C PACKS ON	LAS VEGAS, NV.

C	N1% LMT	01 L TURN	01 R TURN	07 L	07 R	19 L TURN	19 R TURN	CLIMB LIMIT	F
-25	99.10	361.7	382.7	436.0 *	390.6	359.8	346.3	435.6	-13
-15	101.0	355.6	376.7	432.0	388.0	354.9	341.3	435.4	5
-10	101.9	353.0	373.6	428.8	381.5	352.4	338.7	435.4	14
-5	102.8	350.6	370.3	426.2	378.7	350.2	335.7	435.3	23
0	103.7	347.8	366.8	422.8	375.4	347.9	332.9	435.2	32
5	104.7	345.2	362.8	395.4	372.2	345.2	330.2	435.1	41
10	105.6	342.6	359.3	386.9	369.0	343.4	328.3	435.1	50
15	106.5	340.2	356.9	379.3	365.1	340.9	326.7	435.0	59
20	107.5	337.2	354.6	373.0	361.5	338.6	324.9	434.9	68
21	107.6	336.6	354.2	371.4	360.8	338.0	324.8	434.9	70
22	107.8	336.0	353.7	370.1	360.2	337.4	324.5	434.9	72
23	108.0	335.4	353.3	368.6	359.6	336.8	324.1	434.9	73
24	108.2	334.8	352.8	367.1	359.2	336.7	323.7	434.9	75
25	108.4	334.3	352.4	365.7	358.7	336.1	323.4	434.9	77
26	108.6	333.4	351.7	364.0	358.0	335.2	322.9	434.3	79
27	108.7	332.4	351.0	362.2	357.2	334.2	322.3	433.7	81
28	108.7	330.3	349.3	361.6	355.5	332.0	321.0	432.4	82
29	108.6	328.4	346.9	361.3	353.3	329.6	319.4	429.1	84
30	108.4	326.6	344.5	360.3	351.2	327.8	317.7	425.7	86
31	108.2	325.0	342.3	359.5	349.1	326.0	315.7	422.3	88
32	108.1	323.3	340.2	358.8	346.8	324.4	313.7	419.0	90
33	107.9	321.6	337.6	358.2	344.6	322.7	311.9	415.7	91
34	107.7	320.1	335.1	357.9	342.6	321.1	310.1	412.3	93
35	107.6	318.2	332.6	357.3	340.5	319.5	307.9	409.0	95
36	107.4	316.8	330.1	349.9	338.0	317.7	305.7	405.6	97
37	107.2	314.7	328.4	349.9	335.6	315.9	303.6	402.3	99
38	107.1	313.0	326.8	349.9	333.3	314.1	301.7	398.9	100
39	106.9	311.2	325.1	349.9	330.9	312.3	299.7	395.6	102
40	106.7	309.4	323.5	349.9	328.9	310.6	298.0	392.3	104
41	106.6	307.1	321.8	349.9	327.2	308.5	296.3	388.9	106
42	106.4	304.9	320.1	349.9	325.4	306.2	294.6	385.6	108
43	106.2	302.5	318.0	349.9	323.5	303.9	292.7	382.2	109
44	106.1	300.1	315.9	349.9	321.6	301.5	290.9	378.6	111
45	105.7	298.1	313.8	349.9	319.7	299.2	289.4	374.5	113
46	105.4	294.9	310.5	349.9	316.2	296.0	286.0	368.7	115
47	105.0	291.9	306.5	349.9	312.8	293.0	282.8	362.9	117
48	104.7	289.0	303.4	349.9	309.9	291.0	280.7	357.0	118
49	104.3	285.9	300.0	348.8	306.3	288.1	278.1	351.5	120
50	104.0	283.2	297.3	345.5	302.9	285.3	275.7	346.5	122

	01 L	01 R	07 L	07 R	19 L	19 R	CLIMB
RUNWAY LENGTH	8985	9775	14510	10826	9775	8985	-----
RUNWAY SLOPE	-1.1	-1.0	-1.0	-1.0	1.0	1.1	-----
HEADWIND CORR	500	509	498	501	486	479	-----
TAILWIND CORR	-2590	-2665	-6825	-2717	-2496	-2300	-----
KGS/.1" Hg>29.92	0	0	0	0	0	0	742
KGS/.1" Hg<29.92	-3450	-3712	-3616	-3826	-3401	-3315	-1722
LEVEL OFF HT	3181	3181	3181	3181	3181	3181	-----
LIMITED BY	RNWY	RNWY	VMBE	RNWY	RNWY	RNWY	-----

RWY 01LT - ENGINE FAILURE TURN PROCEDURE.
CLIMB RWY HEADING. COMMENCE RIGHT TURN.
AT D3.5 LAS VOR TO 118 DEG MAG HEADING.
INTERCEPT AND TRACK OUTBOUND R-075 LAS VOR.
USE 15 DEGREE BANK ANGLE IN TURN.

RWY 01RT - ENGINE FAILURE TURN PROCEDURE.
CLIMB RWY HEADING. COMMENCE RIGHT TURN.
AT D3.5 LAS VOR TO 118 DEG MAG HEADING.
INTERCEPT AND TRACK OUTBOUND R-075 LAS VOR.
USE 15 DEGREE BANK ANGLE IN TURN.

RWY 19LT - ENGINE FAILURE TURN PROCEDURE.
CLIMB RUNWAY HEADING TO 2.7 DME LAS VOR.
COMMENCE LEFT TURN DIRECT TO LAS VOR.
USE 15 DEG BANK ANGLE IN TURN.

RWY 19RT - ENGINE FAILURE TURN PROCEDURE.
CLIMB RUNWAY HEADING TO 2.7 DME LAS VOR.
COMMENCT LEFT TURN DIRECT TO LAS VOR.
USE 15 DEG BANK ANGLE IN TURN.

ASAP — AIRPORT DATA DATE: 08-Dec-2010

Figure 33: ASAP take-off chart example

4.5 Honeywell

Honeywell International, Inc. is a conglomerate company producing a variety of consumer products, engineering services and aerospace systems. The Global Data Center is one of their businesses, which provide with several flight planning services, weather information, data communication, etc. In May 4th 2010 Honeywell announced the introduction of a runway analyses service available through their website (13).

The Honeywell take-off performance data in a paper form are presented in a similar way as it is done in previous analyses. The heading involves the aircraft type and configuration, airport name and a runway condition, either dry or wet. Based on these, particular analysis has to be selected. The table below heading includes data of the MTOM for up to eight different runways and therefore making it easier to compare individual values for different runways. A code of an appropriate limitation is expressed next to the MTOM value. However it is important to pay attention to the climb limited value of the MTOM, which is stated in the last column. The lower value must be always taken into consideration. It is possible to make a mistake when reading the OAT, because it is expressed in both degrees of Celsius and Fahrenheit. Information about a particular runway length, its slope and acceleration altitude is stated right under each runway. The corrections for a wind and pressure are not directly implied in these analyses and their appropriate values need to be applied when necessary. It is the same for different aircraft system settings and different runway contaminants, which corrections are stated at the table bottom. At the very end of the Honeywell analyses there are described special non-standard departure procedures for engine failure or other relevant notes.

RUNWAY CONDITION: DRY — G550 / G550 BR700-710C4-11 / FLAPS 20 — KASE / ASE ASPEN

ELEVATION= 7820 FT					MAX STRUCTURAL TAKEOFF WEIGHT LIMIT = 91000		
					RUNWAYS - LBS		SEC. SEG. CLIMB
OAT F	OAT C	EPR	BLD ON	A/I ON	15	33THW	
23	-5	1.67	1.66	1.65	0 R	76290 O	91000
25	-4	1.67	1.66	1.65	0 R	76180 O	91000
27	-3	1.67	1.66	1.65	0 R	76070 O	91000
28	-2	1.67	1.66	1.65	0 R	75980 O	91000
30	-1	1.67	1.66	1.65	0 R	75900 O	91000
32	0	1.67	1.66	1.65	0 R	75820 O	91000
34	1	1.67	1.66	1.65	0 R	75730 O	91000
36	2	1.67	1.66	1.65	0 R	75650 O	91000
37	3	1.67	1.66	1.65	0 R	75570 O	91000
39	4	1.67	1.66	1.65	0 R	75500 O	91000
41	5	1.67	1.66	1.66	0 R	75420 O	91000
RUNWAY DIM				LENGTH-FT	0	7006	---
				SLOPE - %	2	-2	---
LVLOFF ALT				FT	0	9320	---
WIND CORR				LBS/KT HW	0	120	---
				LBS/KT TW	N/A	-419	---
QNH				LBS/ 1"Hg > 29.92	0	230	248
				LBS/ 1"Hg < 29.92	N/A	-298	-266
ANTI-ICE				LBS	N/A	-1720	-2750
AC PACKS				LBS	N/A	-960	-1550
SPOILERS				LBS	N/A	-1438	---
ANTI SKID				LBS	N/A	-6343	---

RWY 15
RWY 15 - TAKEOFF NOT PERMITTED.

RWY 33THW - ENGINE FAILURE TURN PROCEDURE..
AT RWY END RIGHT TURN TO 340 DEG MAG HEADING.
LEFT TURN AT D10.2 DBL VOR TO 270 DEG MAG HEADING.
INTERCEPT IPKN LDA NORTHWEST COURSE
USE 15 DEGREE BANK ANGLE IN TURN.
USE 170 KTS MAX SPEED, 125 KTS MIN SPEED IN TURNS

DATE: 05/04/2010 HONEYWELL

Figure 34: Honeywell take-off chart example

4.6 EFRAS

Software EFRAS is a program developed to provide the take-off and landing performance data for several aircraft types including Boeing, Airbus and ATR airplanes. This software was developed and is owned by Condor Flugdienst GmbH, but it is also used by other companies, for example Czech Airlines.

The EFRAS21/TAKEOFF covers both standard and abnormal conditions to calculate the MTOM and v-speeds at brake release for a particular runway. They are expressed as a function of a wind and ambient temperature for all standard aircraft configurations in compliance with all applicable regulations. The parallel calculation of up to four runways is allowed providing the acceleration altitude for each one of them as well. This software is able to consider changes in the aircraft performance for all MEL and CDL items, affects of a runway shortening or different runway contaminants.

Following description of the EFRAS21/TAKEOFF covers the layout of an interface, which was adjusted for the Czech Airlines operation. The input section of this program consists of several cells, where specific information about the indented departure need to be filled. The first cells require inserting the aircraft registration, to obtain the airplane type, and the airport code with a selection of preferred runways. Subsequently, there are inserted the OAT, wind, pressure and runway status, where it is possible to choose from several conditions and contaminants. At the bottom part of the input there are cells, where it is possible to define several aircraft systems settings as well as deviations from a standard configuration. Moreover, the Special input field covers runway modifications or additional performance options, for example runway shortening, adding the obstacle, using the de-icing fluid, etc.

The user can decide to utilize the calculation in two different menus –Main and Details. The performance data provided according to the input data are the same, but the Details menu displays additional background performance information for selected OAT and also three assumed temperatures if applicable. The output section in the Main menu shows the take-off data for selected conditions in easy-to-follow format and consists of the MTOM, v-speeds, limitation code, acceleration altitude, stop margin and other significant information. If a multiple runway calculation is performed, a summary of the calculation is shown in a pop-up window, but the normal output section is displayed after clicking on a particular runway.

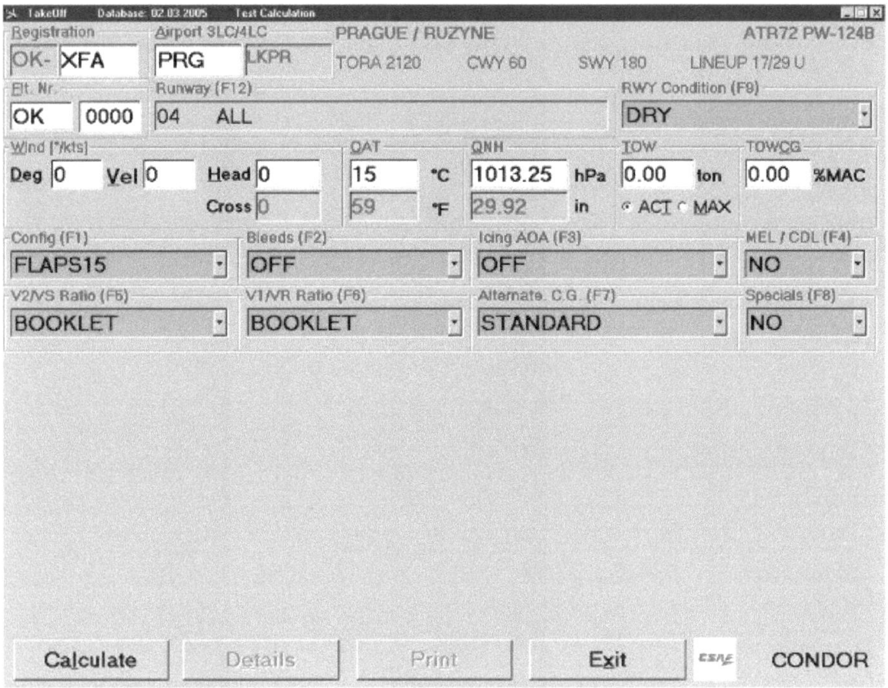

Figure 35: EFRAS input section example

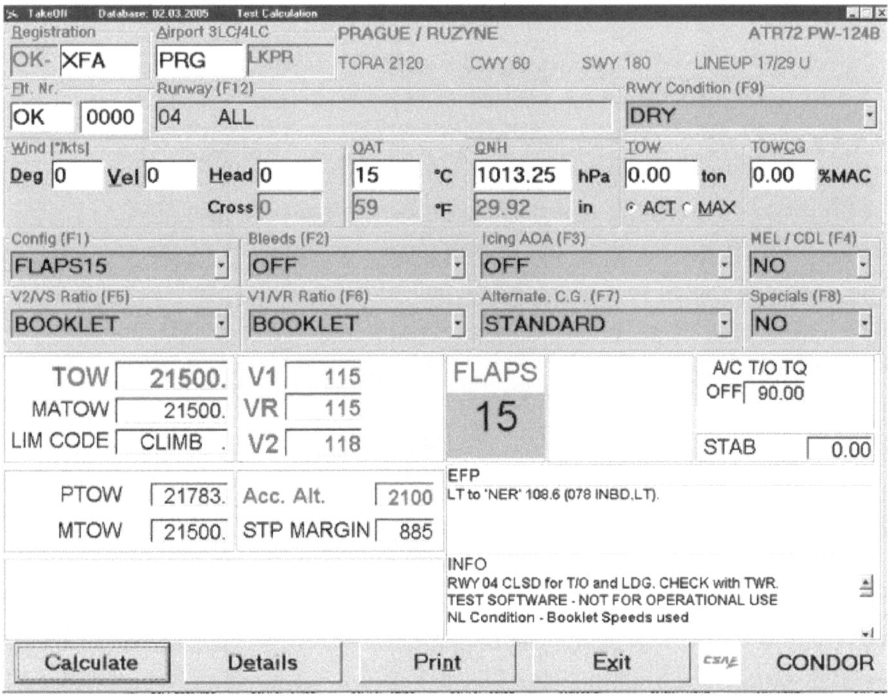

Figure 36: EFRAS Main output menu example

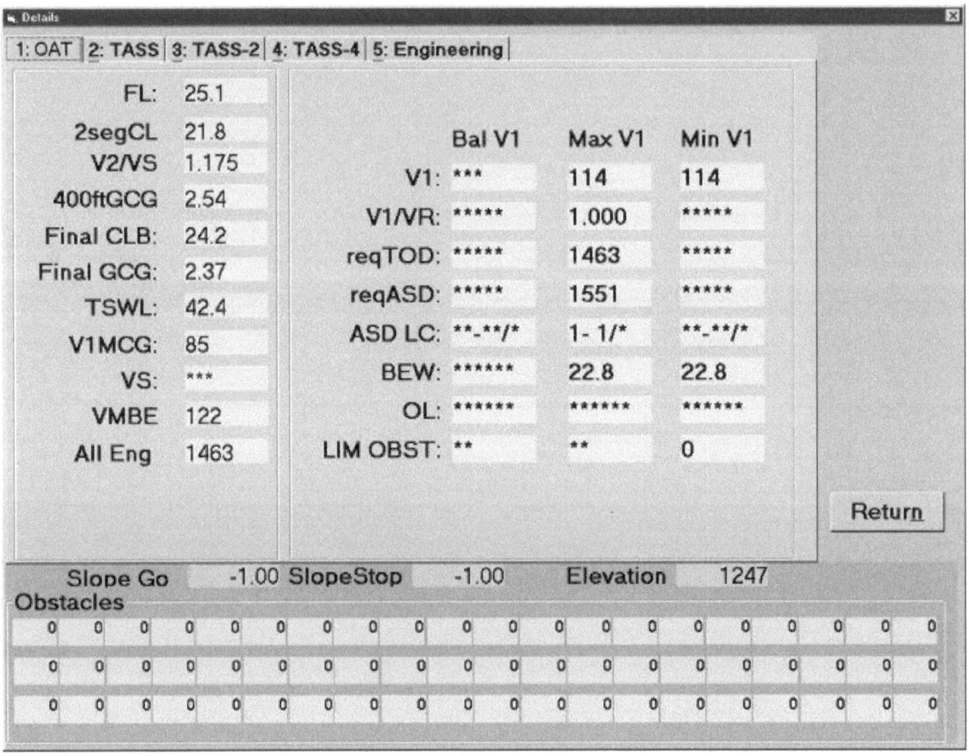

Figure 37: EFRAS Details output menu example

4.7 Conclusion of runway analyses review

At the end of this chapter it is important to mention again that there are many companies with their own products and only a sample of them was described in previous. The sample was selected according to the prestige and a willingness of the companies to provide with exemplary analyses. Nevertheless, all of them should be based on same principles. Different computing programs may however output in different formats.

Based on postulate that the paper form runway analyses should be well-arranged, where it is easy to find the proper value of required take-off data and which minimizes as much as possible the likelihood of errors, probably the most convenient appears the runway analyses format from the Swedish company Flygprestanda. Although they do not provide with parallel data for more runways on one page, these analyses require least additional calculation and therefore reduce the possibility of mistakes. Furthermore, these analyses provide with most information including relevant obstacles height and the brake release point distance. The high quality of the Flygprestanda product was confirmed also by Danube wings pilots and flight dispatchers from the ABS Jets.

Nowadays, almost all companies producing the runway analyses are adjusting their products to offer them also in an electronic form for several platforms. This is based on a wide-spread airlines implementation of the LPC concept. Moreover, these analyses are often accompanied with other pre-flight or in-flight information such as actual weather reports or NOTAMs. This approach results in many advantages, but these systems usually require the Internet connection. The calculation is not then performed at a particular device, but the required data are only downloaded. Therefore, the stand-alone software solutions (for example EFRAS) are irreplaceable in such cases, when there is no chance to connect to the Internet. Moreover, abnormal flight conditions may require a need for a new calculation. This could be performed almost immediately, instead of waiting for the runway analyses provider to do it.

The final decision for a specific runway analyses product depends on a level to which an airline has adopted the LPC concept and which information integrated in modern systems they require. This is always, more or less, limited by financial resources of the particular airline. On the other hand, the companies providing the runway analyses offer their products only for selected airplanes. That is the reason why the decision has to be also based on a supported aircraft types list of a particular company.

5 RUNWAY ANALYSIS CONCEPTUAL MODEL FOR TAKE-OFF

As explained in previous, the optimization process covering all influencing factors is important to operate an airline at the most economical way. Even the regulation allows an aircraft operator to evaluate the TOM only according to the AFM data, almost all airlines use for such purpose software from specialized companies or their own tailored programs. From time to time it may happen that a company providing runway analyses for the airline does not have the data of the MTOM and v-speeds for a particular airport or abnormal aerodrome conditions. The software companies are usually able to perform a new calculation within 24 hours, but especially for a charter operation it may not be soon enough. These are only few reasons why a flight engineering department or even pilots should be able to determine the proper values of the MTOM and appropriate v-speeds only according to the AFM data or Flight Crew Operating Manual (FCOM).

The objective of this section is to cover a whole process of obtaining the runway analysis data for the take-off only according to the AFM data provided by the aircraft manufacturer. The process of this calculation method is described step-by-step by means of flowcharts. Moreover, this conceptual model may also serve as a heart of calculation software, which would be able to evaluate the take-off data automatically.

5.1 Input data

First of all, to obtain required values of the MTOM and v-speeds it is essential to know all local and atmospheric conditions. A collection of these represents the Input data for the take-off calculation. The following table presents all necessary data to perform it.

Table 6: Input data for take-off calculation

Input data	
runway	TORA TODA ASDA elevation slope contamination
atmospheric conditions	OAT wind pressure
obstacles	height distance
aircraft	status configuration

An aircraft configuration, although it is a free parameter which could significantly affect the final take-off performance, need to be considered for this calculation purpose as fixed. Therefore, for each individual setting an independent calculation need to be performed. Moreover, it is important to be aware that all available aerodrome distances (TORA, TODA, and ASDA) need to be reduced by the distance of the aircraft line-up before their entering into the calculation process.

5.2 Mass to v_1/v_r ratio graph

After acquiring all Input data, next step is to design an initial "Mass versus v_1/v_r ratio" graph, which will be later subjected to the optimization. The process to obtain such graph is translated by flowchart illustrated below.

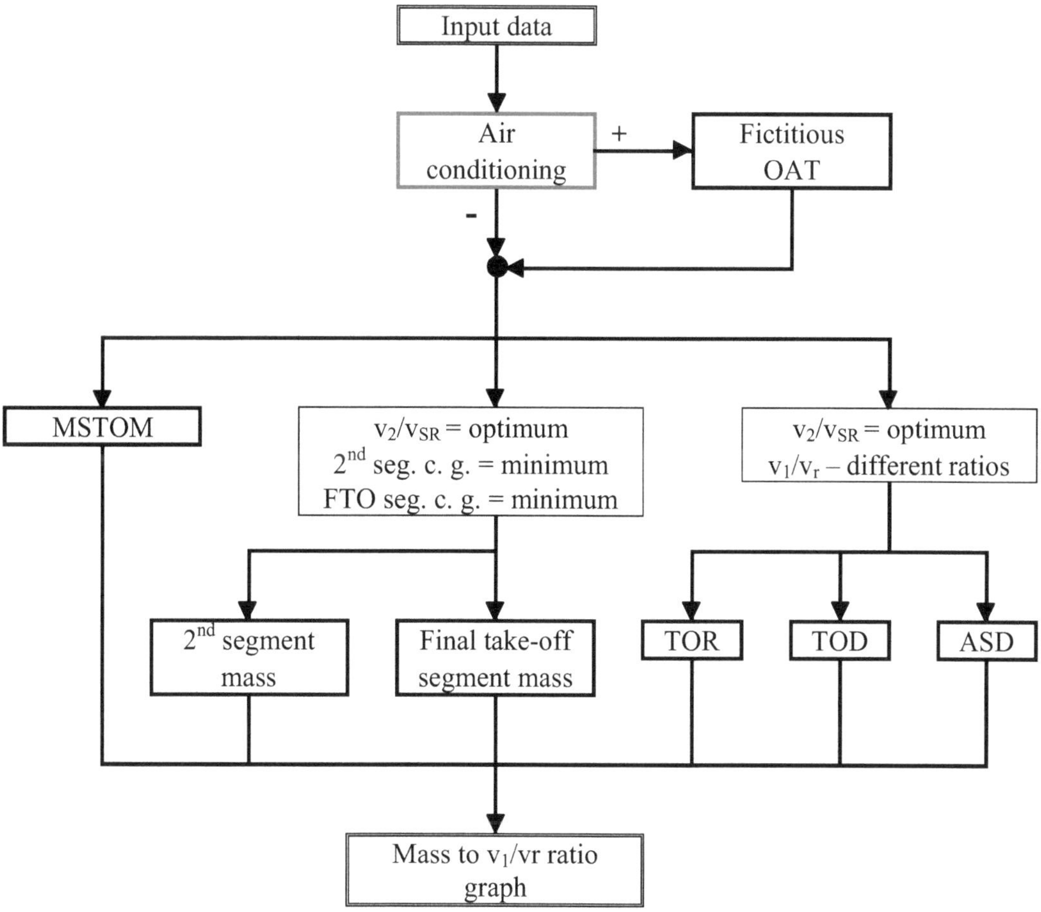

Figure 38: Mass to v_1/v_r ratio graph flowchart

An initial and final cell of the flowchart is designed in a double-line border to emphasize the beginning and the end of a particular part of the calculation process. Cells with a blue border represent decision process to determine if the statement is true or false. Thick border symbolizes that a particular calculation according to the AFM data needs to be performed to obtain desired data. On the other hand, thin border just indicates that specific values of some parameters need to be applied.

The figure 38 represents the flowchart to obtain the "Mass versus v_1/v_r ratio" graph. The optimum value of v_2/v_{SR} in this process means the value to be inserted, which allows highest masses in a second segment gradient limitation. It is usually the highest manufacturer stated value of the v_2/v_{SR} range, but it may not be always true. When computing the TOM limited by the 2nd segment and the final take-off segment climb gradient, according to the number of engines initially enter minimal values required by the regulation, because they provide with highest masses.

When determining various masses limited by the TOR, TOD and ASD for different v_1/v_r ratios, it is necessary to input relevant values of the TORA, TODA and ASDA reduced by the distance of the aircraft line-up. In addition to that, in case of icing conditions or a non dry, contaminated runway, appropriate corrections or particular data need to be applied. This is because these conditions influence either climb performance or aircraft acceleration phase on the ground.

5.3 Optimization process

After designing the initial "Mass versus v_1/v_r ratio" graph, it is possible to evaluate an accurate value of the MTOM and appropriate speed ratios by means of the optimization principle described in chapter 3.

First of all, the MTOM need to be selected from the initial "Mass versus v_1/v_r ratio" graph. This mass could correspond to exactly one value of the v_1/v_r ratio or a specific range. Nevertheless, it is always advisable to select the highest v_1/v_r ratio for the particular MTOM as it provides with a shortest TOD and later with take-off speeds with a highest margin to v_{MCG} (minimum control speed on ground). After that, for these particular ratios of v_1/v_r, v_2/v_{SR} and selected MTOM, it is necessary to calculate required TOD to be able to locate all relevant obstacles according to the point, where the flight phase of the take-off begins.

The aim of the optimization process is to find an optimum v_2/v_{SR} ratio. In previous step this ratio was set to the value, which provides with the highest TOM limited by the 2nd segment climb gradient. Consequently, if after a study of the "Mass versus v_1/v_r ratio" graph we estimate that the MTOM is limited by the climb gradients, no changes in v_2/v_{SR} ratio are necessary. It is similar to the MTOM limited by the MSTOM as it is independent from any ratio change. On the other hand, all other limitations are influenced by the v_2/v_{SR} ratio. The lower the ratio is the less restrictive these limitations are. Therefore, if the MTOM is limited by the available runway distances only, lowering the v_2/v_{SR} ratio may and continue with a new graph designed according to the new v_2/v_{SR} ratio.

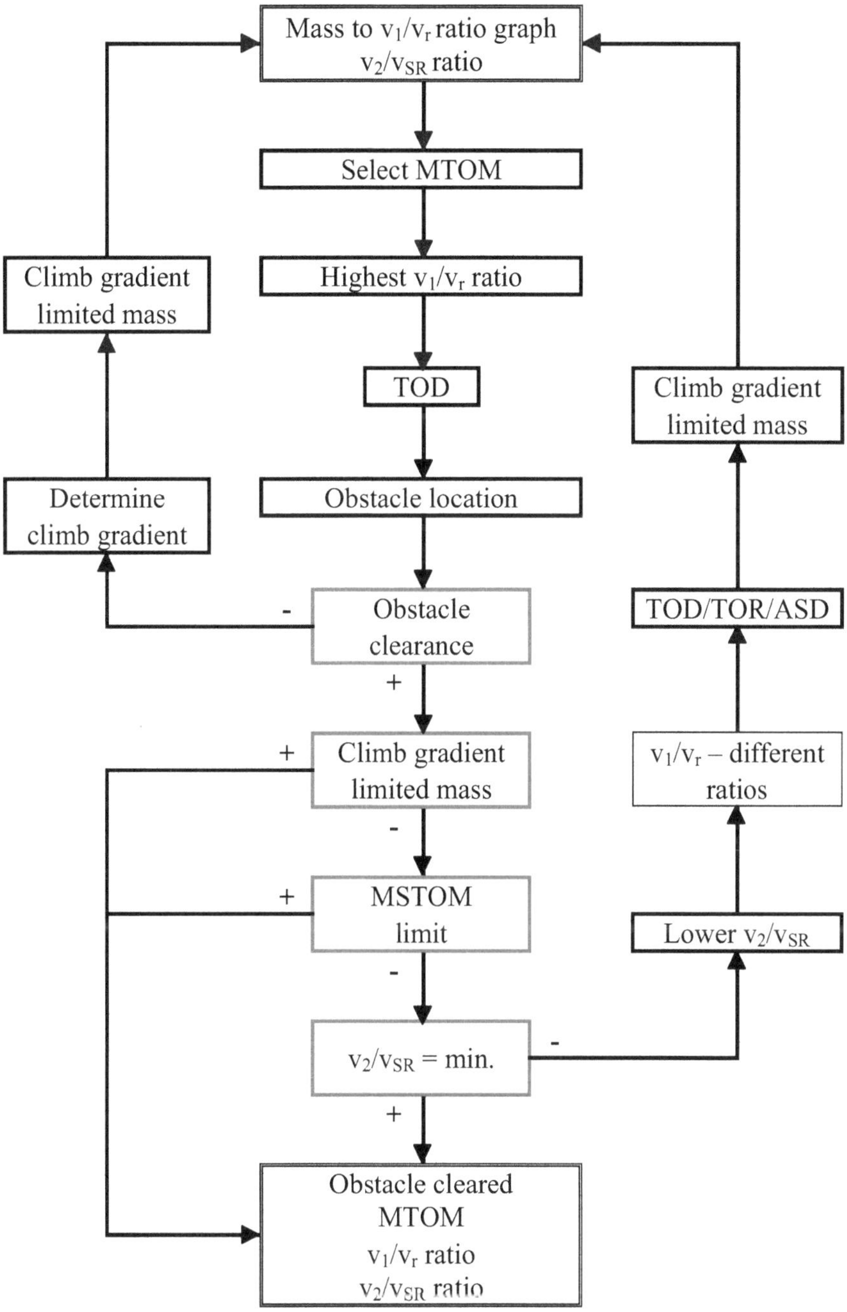

Figure 39: Optimization process flowchart

In any case, it is important to always check for an obstacle clearance. If the aircraft does not clear any of them, it is obligatory to figure out a necessary climb gradient and a new limiting mass. Only after proving that the TOM along with appropriate climb gradients allows all relevant obstacles clearance this particular mass could be declared as the MTOM.

It may happen that during the calculation the MTOM may fall below the value of the operating mass. This mass represents a sum of the aircraft, pantry, crew and their baggage, all operating liquids and useable fuel. Consequently, the take-off is not possible at these conditions and the calculation could be ended.

5.4 v-speeds evaluation

Last section of the take-off data calculation process is an evaluation of the take-off speeds. The beginning of this particular process requires that all relevant obstacles are cleared and the values of the MTOM, the appropriate v_1/v_r and v_2/v_{SR} ratio are known. As mentioned in a very beginning of the chapter 2, the values of the MTOM and appropriate v-speeds go hand in hand and therefore changes in both the MTOM and the speed ratios may still occur.

First step of the evaluation is to determine the rotation speed v_r for the particular MTOM and consequently the decision speed v_1 from known v_1/v_r ratio. After that, these speeds are checked to meet the regulation requirements for v_{MCG}, respectively v_{MCA}. Moreover, it must be verified that the decision speed v_1 is not higher than a computed decision speed limited by the brake energy. If all requirements for these speeds are fulfilled, v_2 (take-off safety climb speed with a critical engine failure) could be estimated. This needs to be checked to be at least $1{,}1v_{MCA}$ required by the regulation. If confirmed, the process of obtaining the MTOM and appropriate v-speeds for the particular runway and environment conditions is finally at the end.

In case that any v-speeds limitation is not satisfied, the flowchart illustrated in the Figure 28 translates the actions to be followed. This means a change in either the v_1/v_r or v_2/v_{SR} ratio. The adjustment in the v_2/v_{SR} ratio supposes a performing of a new calculation to obtain a new "Mass versus v_1/v_r ratio" graph designed according to the new value of this ratio. Lowering of the v_1/v_r ratio does not necessarily mean a reduction of the MTOM, but the lower value of this ratio prolongs the TOD. Consequently, the obstacle distance from the point, where the flight phase of the take-off begins, is reduced. This may lead to stricter climb gradients. Nevertheless, in all cases it is required to select a new MTOM and restart the v-speeds calculation process.

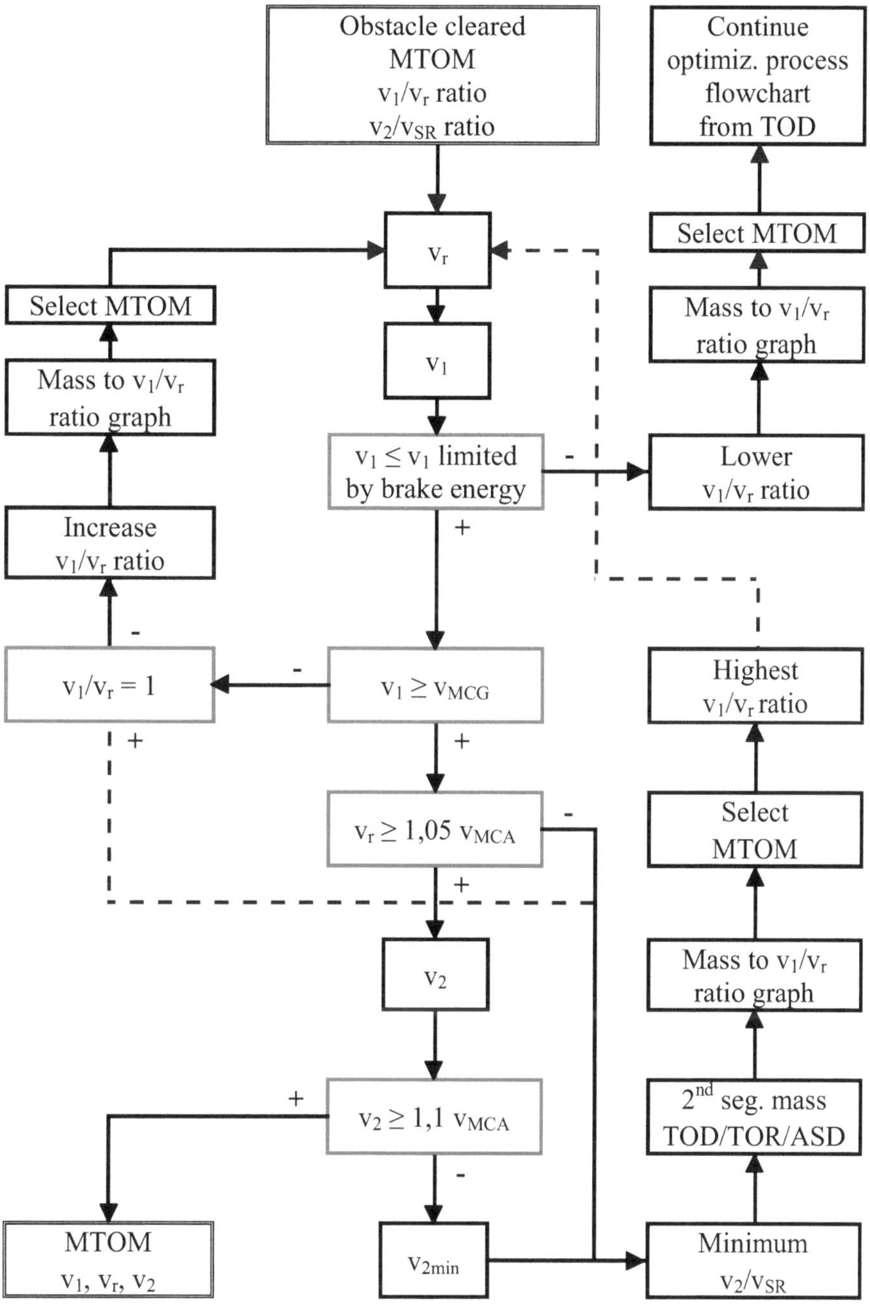

Figure 40: v-speeds evaluation flowchart

CONCLUSION

A personal experience of the author with using the runway analyses tables raised a curiosity which all factors may contribute in computing of the take-off data. Moreover, it was a challenge to create a conceptual model of the calculation process to be able to compute the appropriate values of the MTOM and v-speeds for various conditions. These few were main reasons to choose this matter as a subject of research.

Understanding the way in which the influencing factors affect the take-off is the first essential condition to perform the calculation. This is a reason why there was paid a great attention to sum up all of them. This section could also serve as a study material, where all influencing factors are described and thus fulfills the first objective of this book.

The aircraft performance data calculation reflects through the whole airline operation mainly due to economic and safety reasons. A flight engineering department or pilots may be required to do the calculation for current ambient conditions only according to the performance data included in the AFM. This is the reason why all of them should understand at least basic principles of this calculation. Furthermore, the optimization process with a sufficient precision might be a key to gain a competitive advantage. A particular section of this paper offers a description of the optimization process principle according to the currently valid regulation. Consequently, there was created the conceptual model of the take-off data calculation in a form of flowcharts. Providing that all necessary performance data are available, all relevant personnel should be able to perform the calculation and obtain the proper values of the MTOM and v-speeds for specific aerodrome and flight conditions by following of these flowcharts. An example of such calculation according to the created flowcharts could be found in an Appendix..

Designing of the flowcharts supposes the creation of a specialized software application. This would gain the take-off data from particular aircraft performance characteristics and therefore reduce the time required to evaluate all influents manually. Although this is the objective for programmers and currently it is not in author's capability, the conceptual model should be a backbone of this software application. The chapter 4 of this book compares the selected runway analyses products trying to identify both the advantages and drawbacks of each one of them. This was important because it provides with information which design and form is the most advisable to use, respectively which positives should be adopted in case of programming the application.

Finally, the author hopes that information included in this book sufficiently answer all people, who are similarly interested in runway analyses.

BIBLIOGRAPHY

1. **Ojha, S.K.** *Flight Performance of Aircraft.* American Institute of Aeronatics and Astronautics, 1995. ISBN: 1-56347-113-2.
2. EASA extension of scope: timetable. [Online] [Cited: April 12, 2012.] http://easa.europa.eu/flightstandards/npa.html.
3. **European Commission.** *Commission regulation (EC) No 859/2008.* 20 August 2008.
4. **Chmelík, J.** Plánování a provádění letu. In Volner R. et al. *Flight planning management.* CERM, 2007. ISBN: 978-80-7204-496-2.
5. **Honeywel GDC.** *Runway analysis Aircraft performance data.* 2010.
6. **ICAO.** *Aerodrome Design Manual.* 3. 2006. Vol. 1. ISBN 878-92-9231-065-3.
7. **AIRBUS.** *Getting to grips with aicraft performance.* 2002.
8. **European Aviation Safety Agency.** *Certification specifications and acceptable means of compliance for large aeroplanes CS-25.* 2011.
9. **European Commission.** *Acceptable means of compliance and interpretative.* IEM OPS 1.495(a) Take-off obstacle clearance. 2001.
10. **Kazda, A. and Caves, R.E.** *Airport design and operation.* Elsevier, 2007. ISBN: 978-0-08-045104-6.
11. **Viano, J.P. R.d.L.** *Takeoff and Landing Performance Optimization.* Master thesis : Universidade Tecnica de Lisboa, 2011.
12. **Aguiar, A.** *Aeroplane Performance.* TAP, 2005.
13. Honeywell Introduces Runway Analysis. [Online] May 4, 2010. [Cited: February 2, 2012.] http://www51.honeywell.com/honeywell/news-events/press-releases-details/05.04.10_RunwayAnalysis.html.
14. **Jeppessen Sanderson inc.** *JAA ATPL Training. Meteorology.* Jeppessen GmbH, 2001. ISBN 0-88487-285-8.
15. **Krollová, S., Jůn, F. and Topoľčány, R.** *Meteorológia.* EDIS. ISBN: 80-8070-216-0.
16. **Volner, R. et al.** *Flight Planning Management.* CERM, 2007. ISBN: 80-7204-496-2.
17. **Jeppessen Sanderson inc.** *JAA ATPL Training. Airframes and systems.* Jeppessen GmbH, 2001. ISBN: 0-88487-285-8.
18. **ASAP.** *Runway Analysis Description.* 2011.

19. **ATR.** *ATR 72 Airplane Flight Manual. Revision 24.* 2010.

20. **ICAO.** *Annex 6 Operation of Aircraft.* 2001.

21. **Condor Flugdienst GmbH.** *EFRAS21/TAKEOFF MENU DESCRIPTION.* 2005.

22. **Jeppessen Sanderson inc.** *JAA ATPL Trainning. Mass & Balance and Performance.* Jeppessen GmbH, 2001. ISBN: 0-88487-285-8.

23. **Bugaj, M. and Januš, J.** *Letové výkony a plánovanie. Hmotnosť a vyváženie. Výkony.* EDIS. ISBN: 80-8070-213-6.

24. **ATR.** *ATR 72 Flight Crew Operating Manual.* 1999.

25. **Airbus.** *A318/A319/A320/A321 Flight Crew Operating Manual, Flight preparation.*

26. **Hartmann, P.V.** *2 nebo 4.* [Online] January 25, 2012 [Cited: March 2, 2012] http://www.planes.cz/cs/clanky/508/

27. **Kříž, J.** *Pohonná jednotka, JAA ATPL výcvik.* EDIS, 2008. ISBN: 978-80-8070-8726.

28. **Soldán, V.** *Letové postupy a provoz letadel.* 2007. ISBN: 978-80-239-8595-5.

APPENDIX

This appendix to book *Runway Analyses Application for Take-off* provides with an example of the take-off data calculation for a specific aircraft type, aerodrome and ambient conditions. The computing is performed according to the created flowcharts.

For this example an aircraft ATR 72-200 and airport Košice were chosen. The source of the aircraft performance data represents the AFM performance graphs, which could be found in chapter Performance, subsection Take-off. The aerodrome data were obtained from the AIP Slovak republic. Local ambient weather conditions were imagined.

LZKZ – Košice

Runway 01

True bearing	013 degrees
Elevation	755 ft
Runway slope	+ 0,2%
TORA	3 100 meters
TODA	3 300 meters
ASDA	3 100 meters
Line-up correction	90 degrees

Runway 19

True bearing	193 degrees
Elevation	755 ft
Runway slope	- 0,2%
TORA	3 100 meters
TODA	3 300 meters
ASDA	3 100 meters
Line-up correction	90 degrees

The Košice airport is equipped with one runway 01/19. The objective of this calculation is also to decide which runway direction is more convenient for the take-off in term of the MTOM.

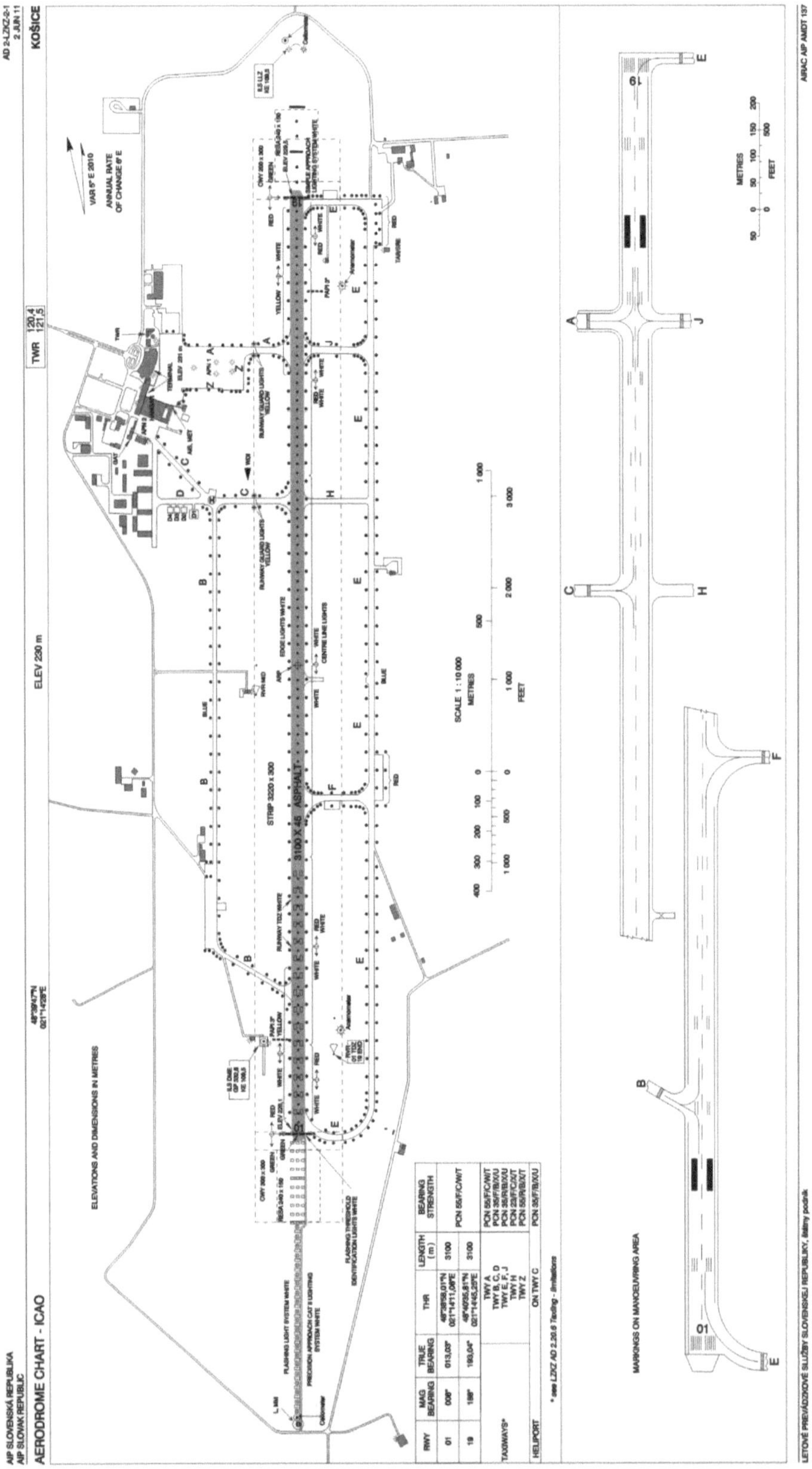

Figure A1: LZKZ aerodrome chart [Source: AIP Slovak republic]

Ambient weather conditions

METAR LZKZ 211530Z 32515KT 9999 SCT040 25/21 Q1008 NOSIG=

Wind direction	325 degrees (true)
Wind velocity	15 knots
Relative rwy 01 wind	10 knots head
Relative rwy 19 wind	10 knots tail
Ambient temperature	25 degrees of Celsius
Ambient pressure	1008 hPa
Contaminants	None
Airport pressure altitude	895 ft

ATR 72-200

A/C status	OK
Air conditioning	OFF
TORA 90 degrees line-up correction	11 meters
TODA 90 degrees line-up correction	11 meters
ASDA 90 degrees line-up correction	24 meters

rwy 01 90 degrees line-up correction	
TORA	3 089 meters
TODA	3 289 meters
ASDA	3 076 meters

rwy 19 90 degrees line-up correction	
TORA	3 089 meters
TODA	3 289 meters
ASDA	3 076 meters

Mass to v1/vr ratio graph

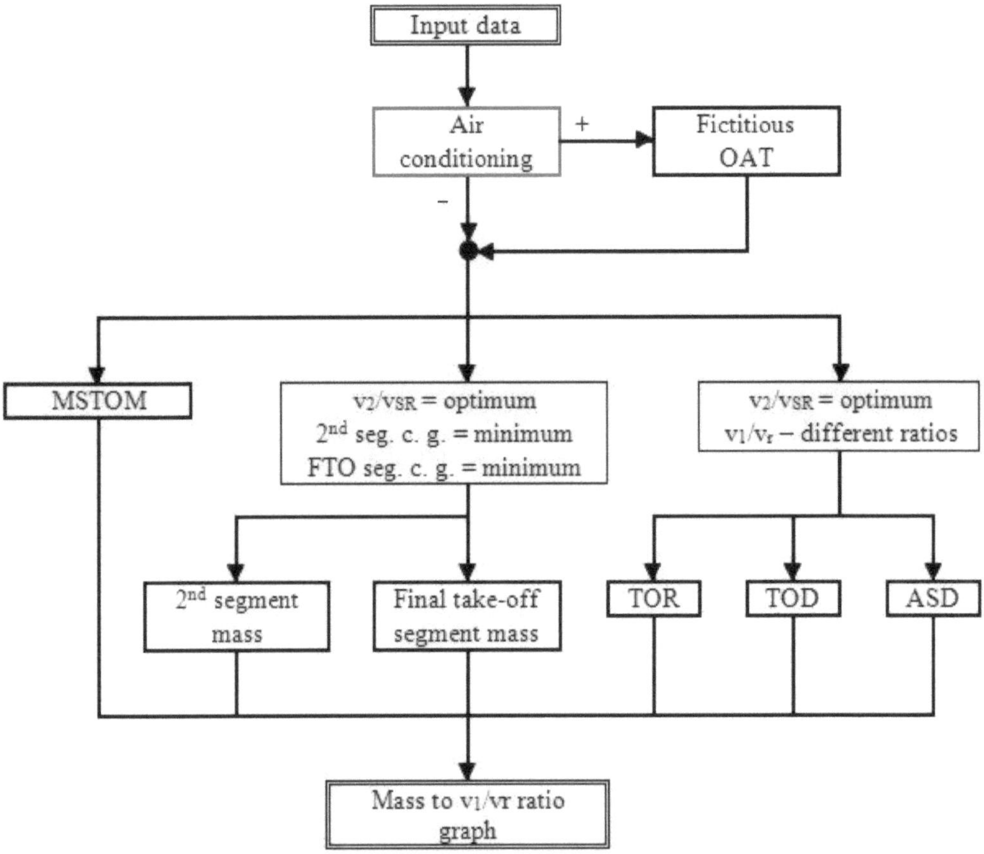

Figure A2: Mass to v1/vr ratio graph flowchart

v_2/v_{SR} ratio	1,18
2nd segment climb gradient	2,4
FTO segment climb gradient	1,2

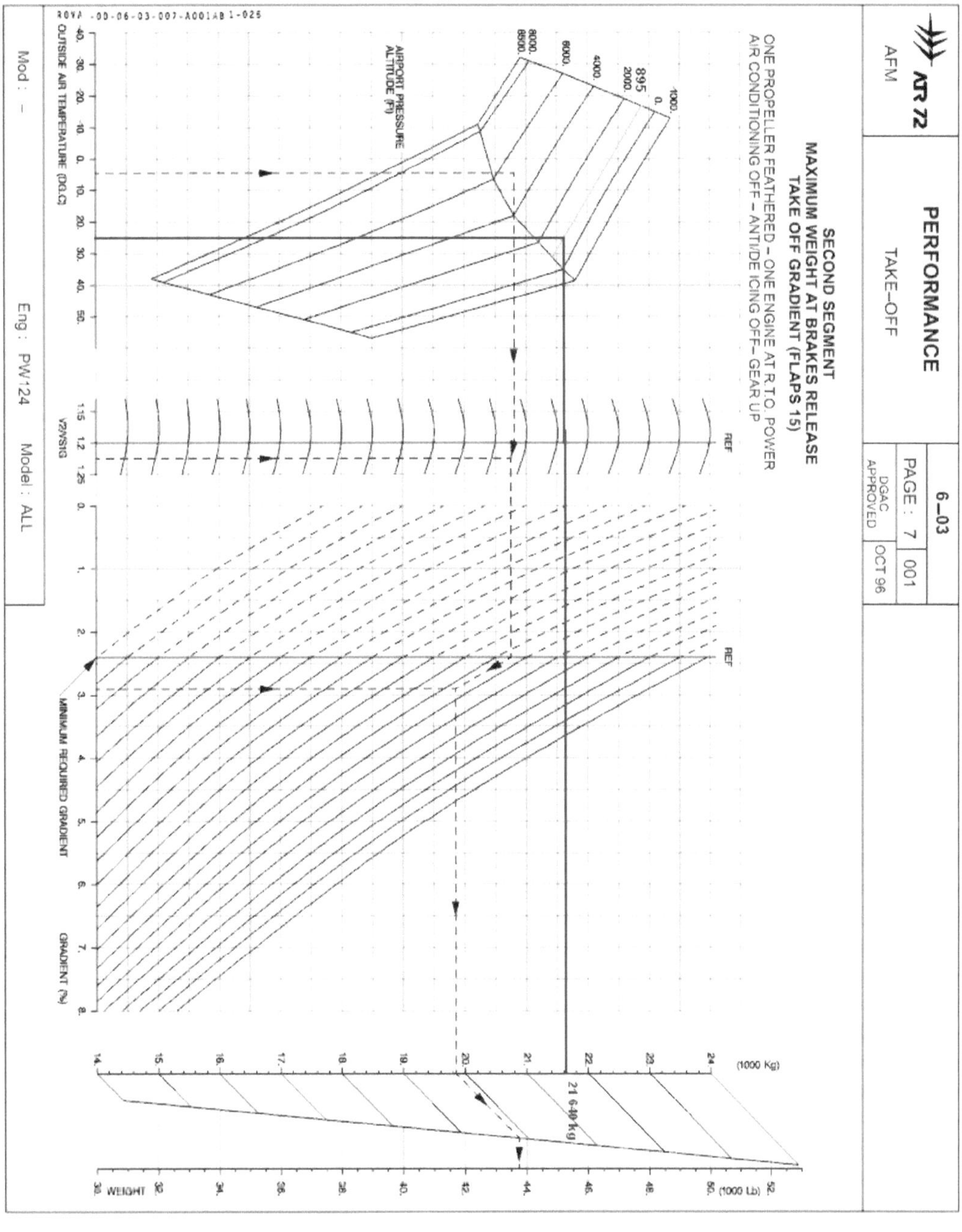

Figure A3: 2nd segment climb gradient limited mass [Source: ATR, ATR 72 AFM]

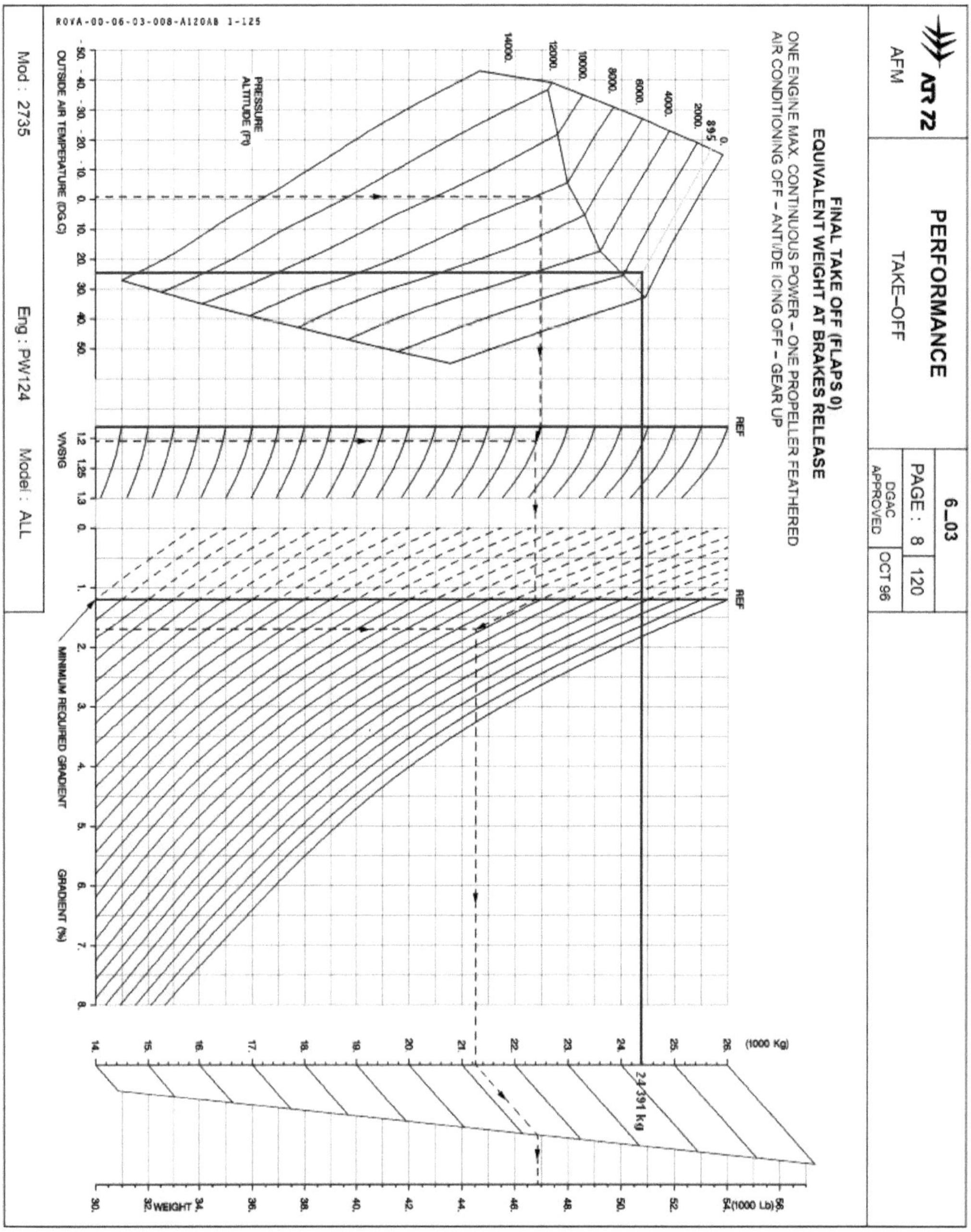

Figure A4: Final take-off segment climb gradient limited mass [Source: ATR, ATR 72 AFM]

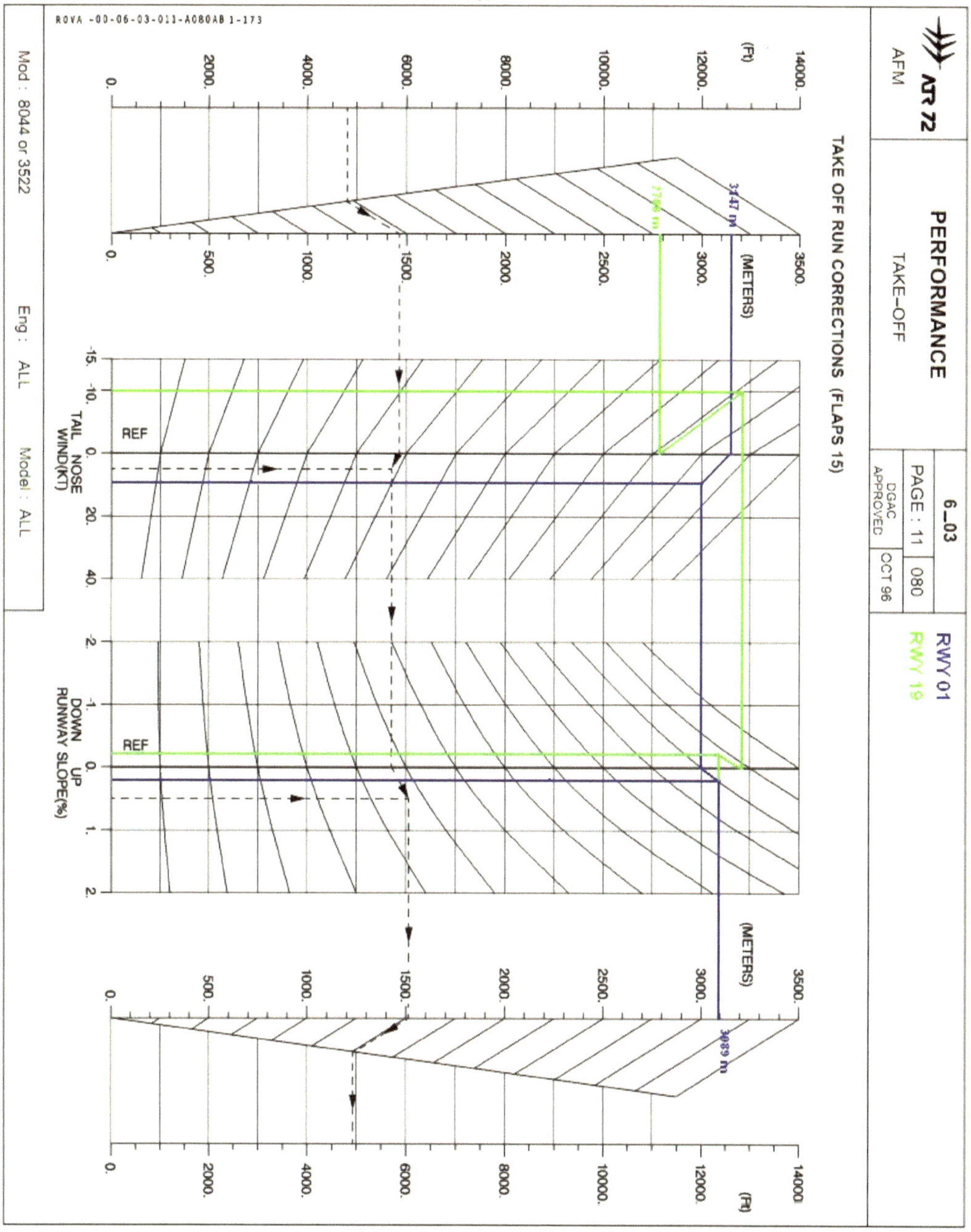

Figure A5: TOR corrections [Source: ATR, ATR 72 AFM]

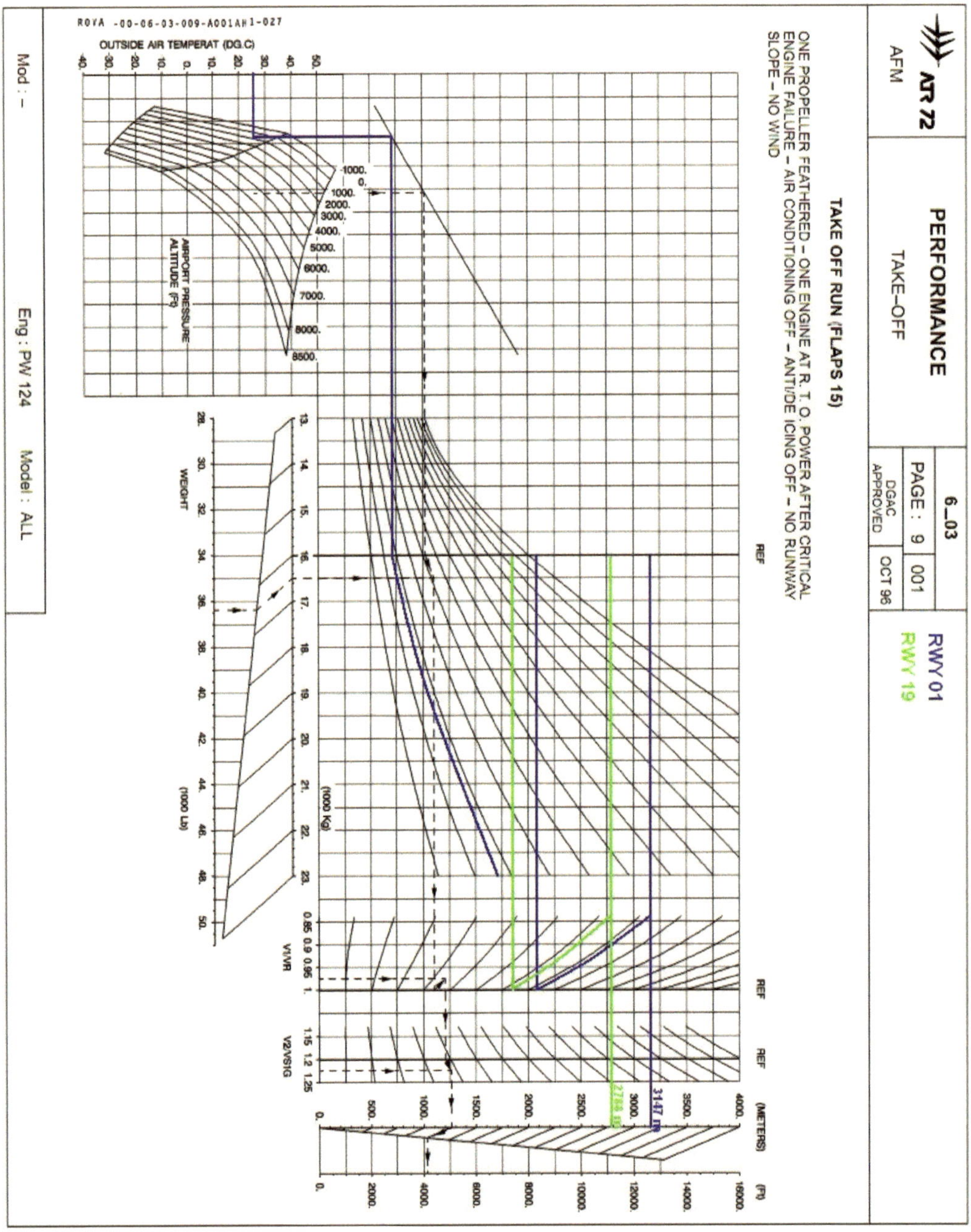

Figure A6: TOR 1 engine out limited mass [Source: ATR, ATR 72 AFM]

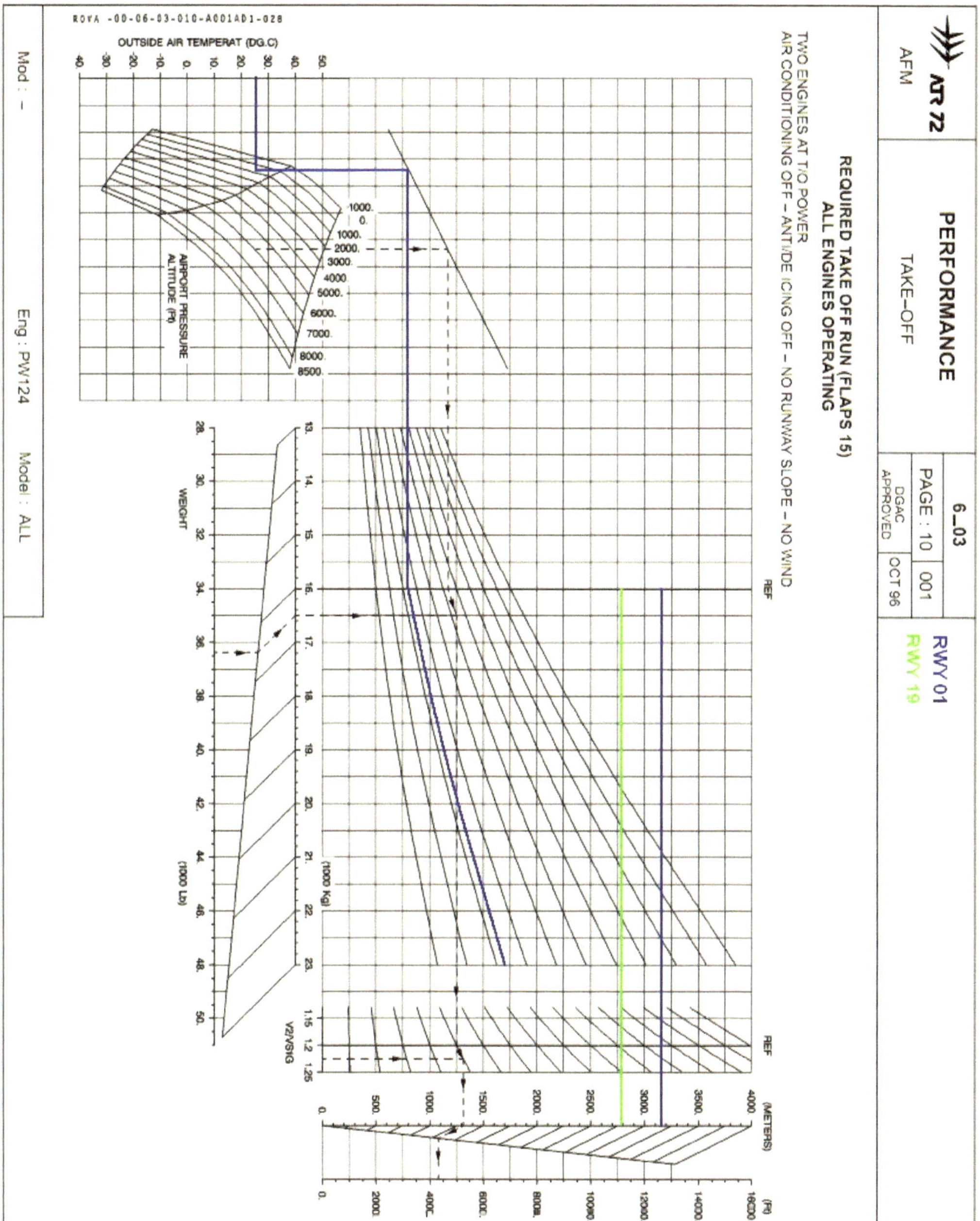

Figure A7: TOR all engines limited mass [Source: ATR, ATR 72 AFM]

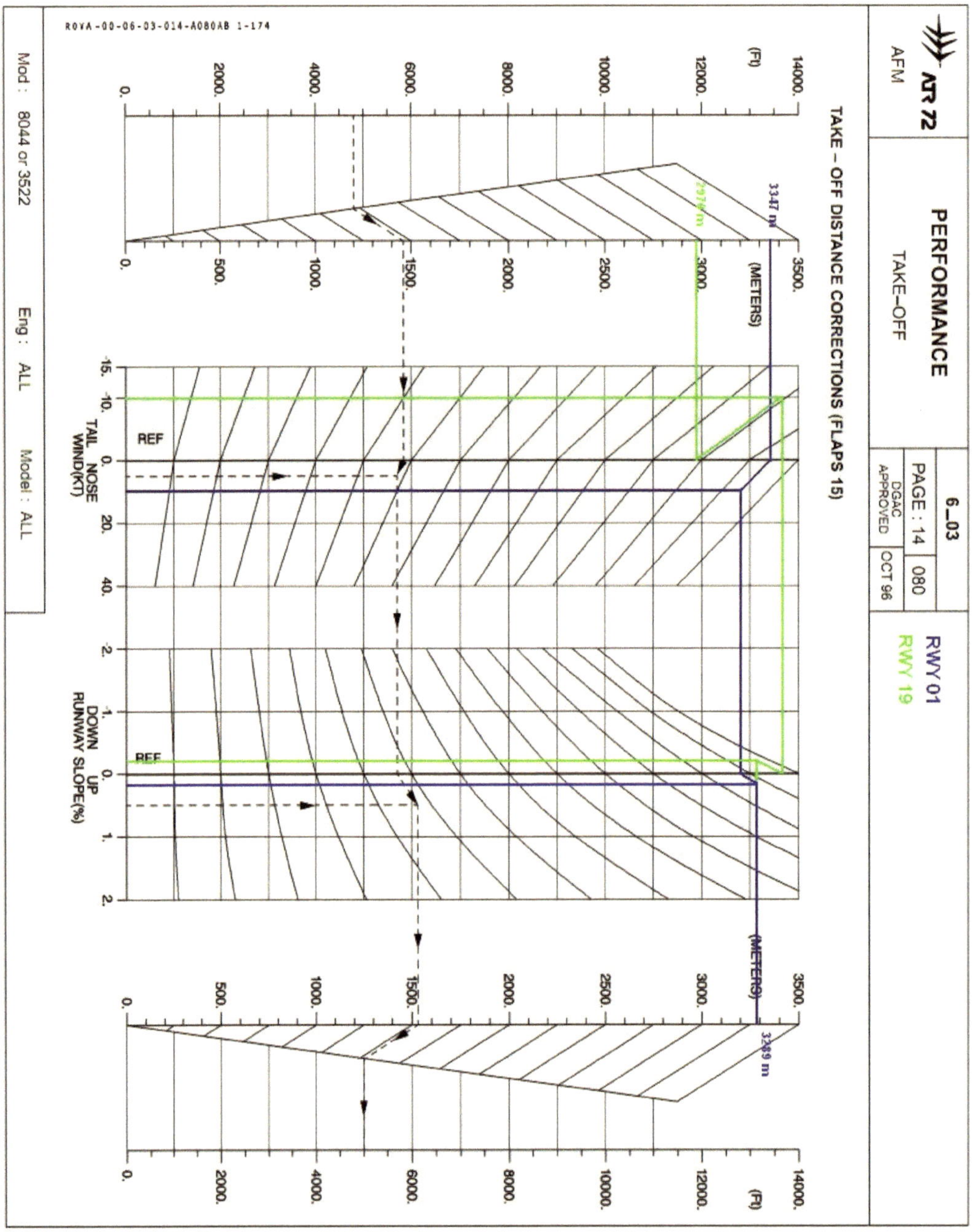

Figure A8: TOD corrections [Source: ATR, ATR 72 AFM]

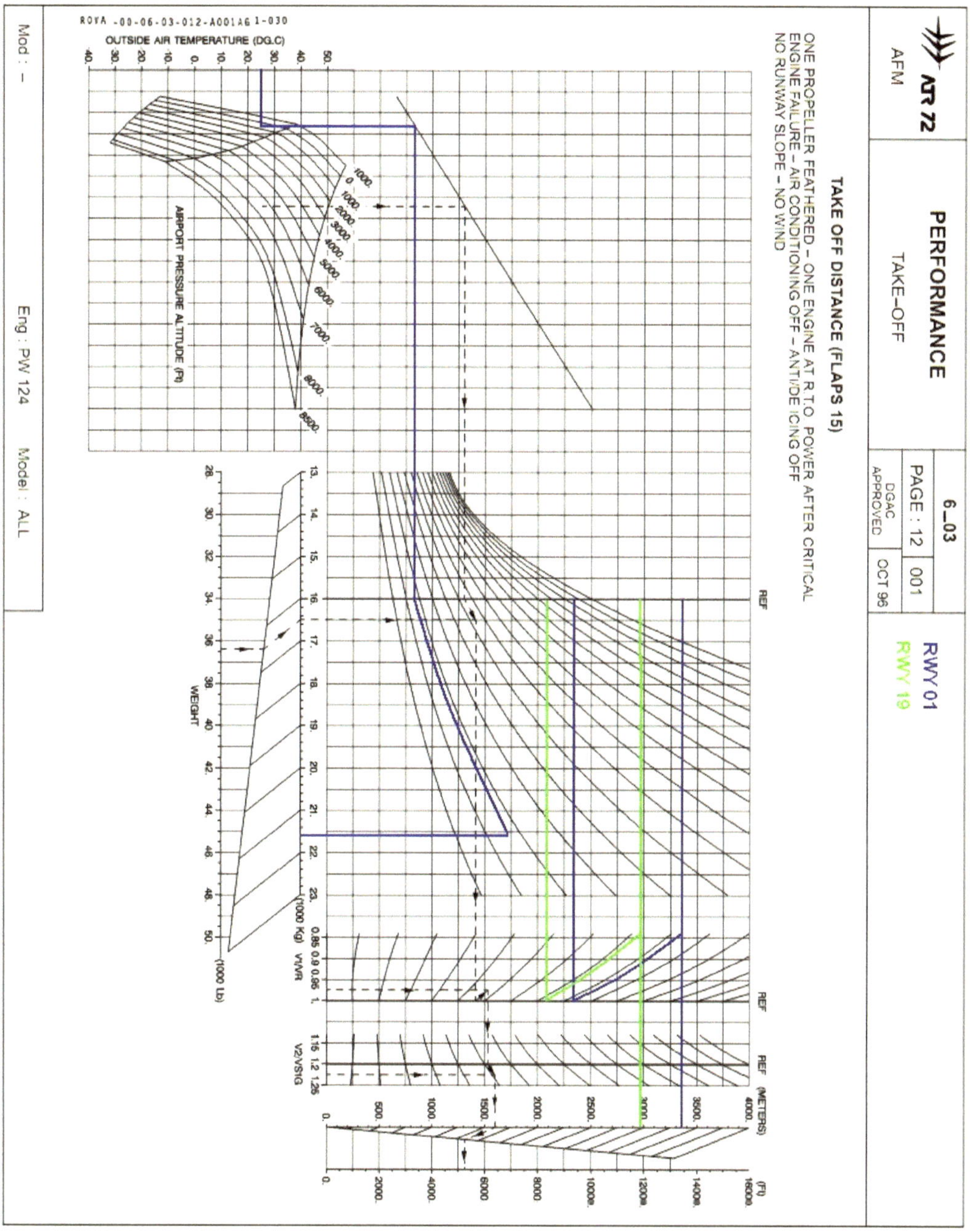

Figure A9: TOD 1 engine out limited mass [Source: ATR, ATR 72 AFM]

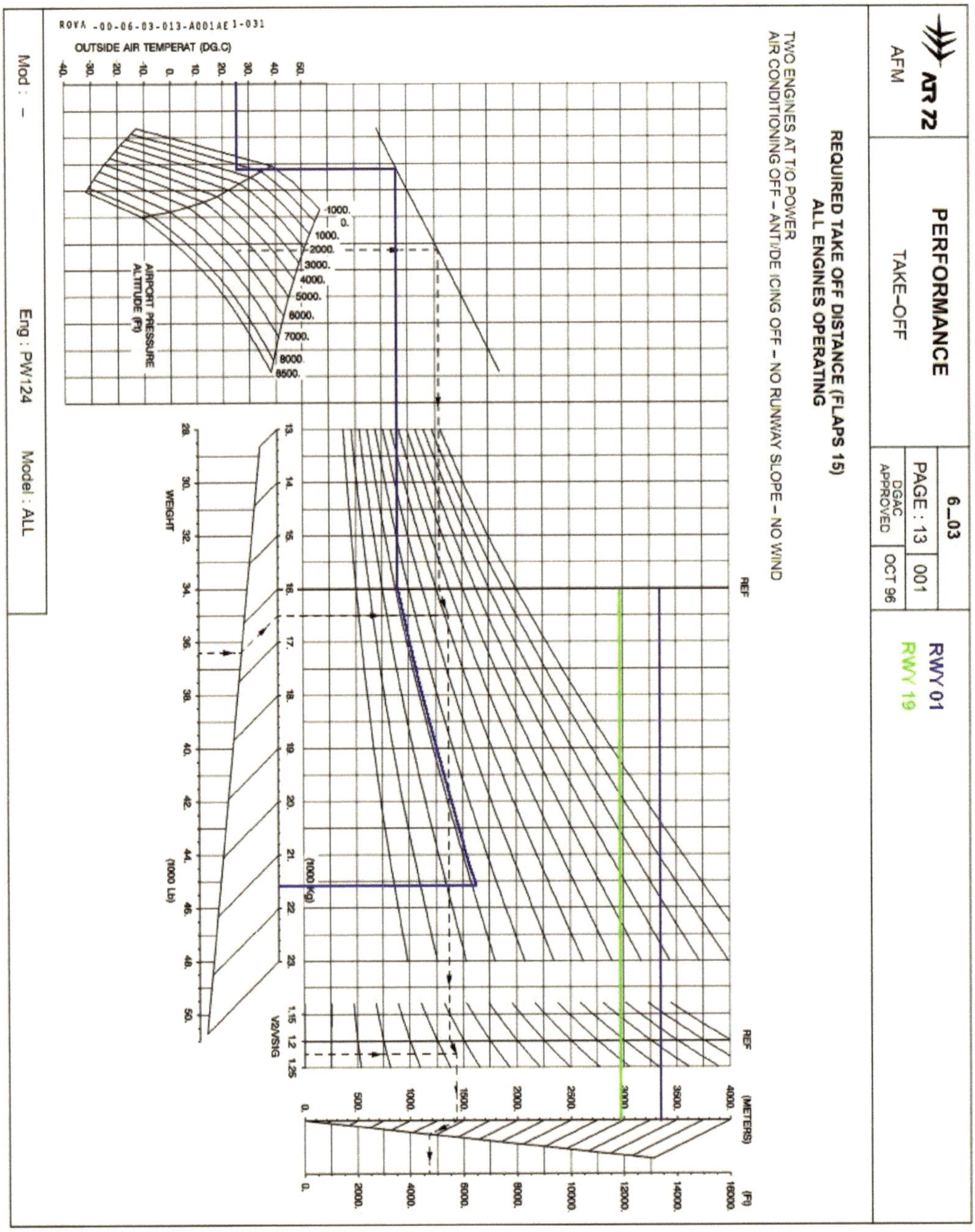

Figure A10: TOD all engines limited mass [Source: ATR, ATR 72 AFM]

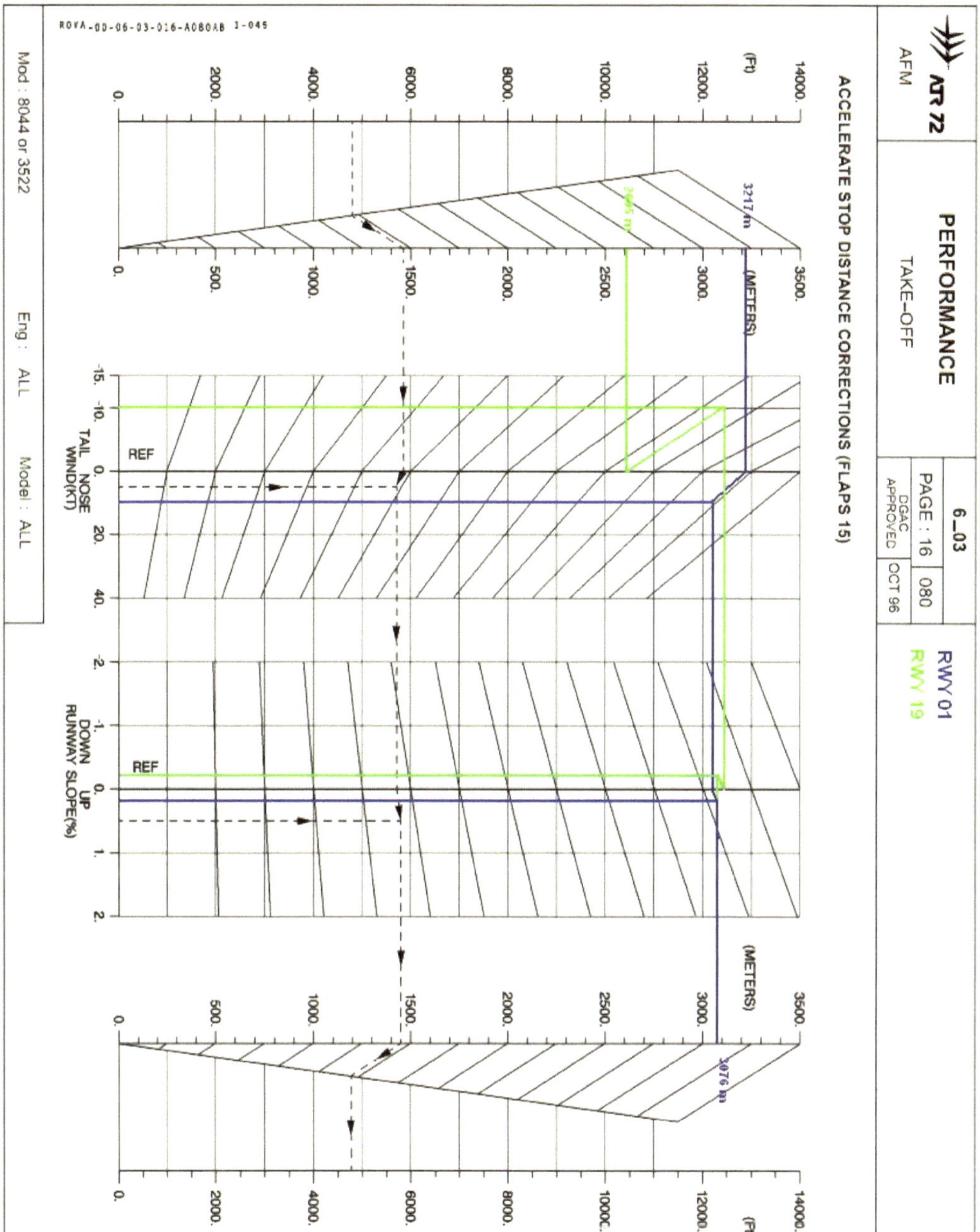

Figure A11: ASD corrections [Source: ATR, ATR 72 AFM]

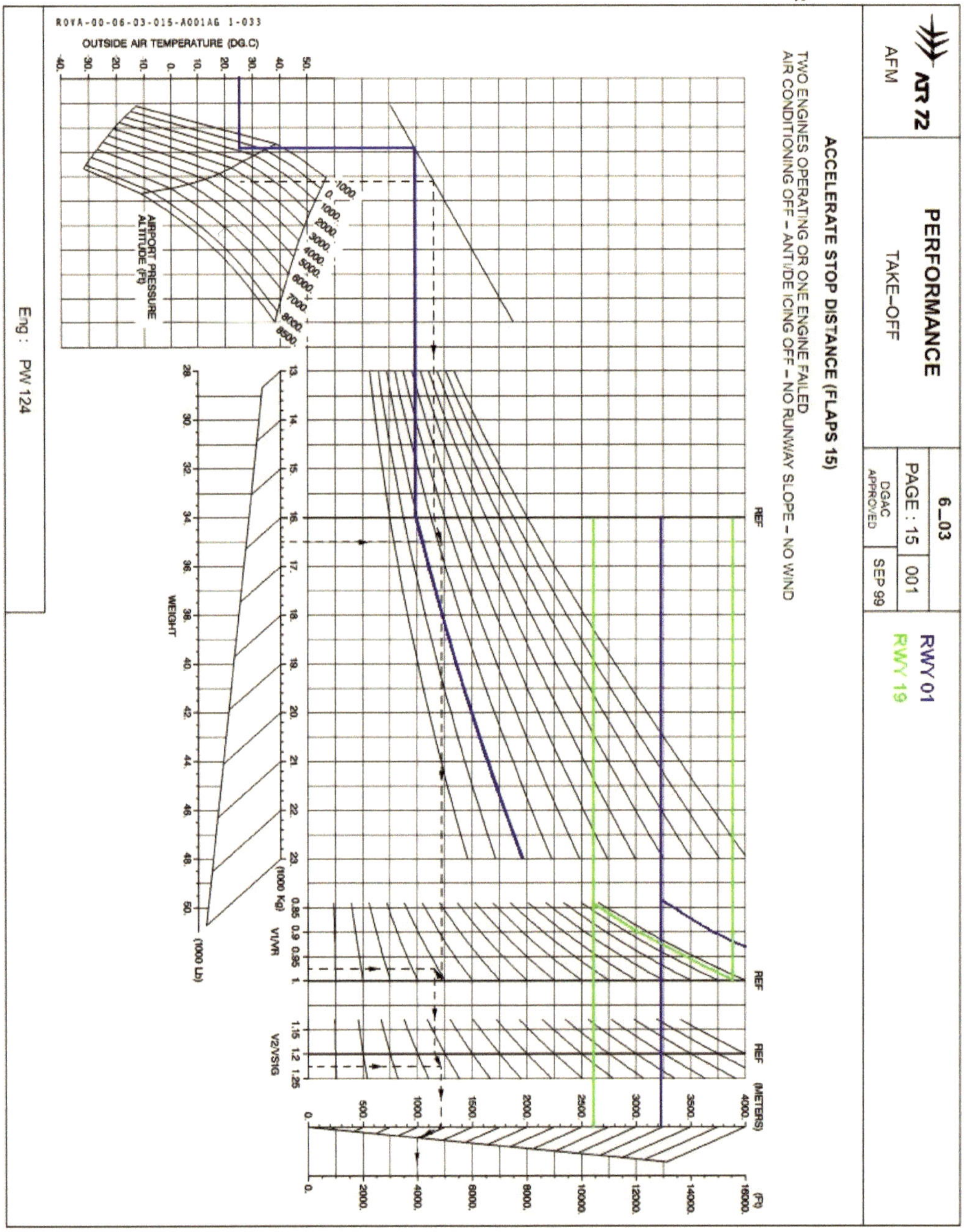

Figure A12: ASD limited mass [Source: ATR, ATR 72 AFM]

LZKZ RWY 01

LZKZ RWY 01		v1/vr	
v2/vsr = 1,18		0,84	1
mass [kg]	MSTOM	22 000	22 000
	2nd segment	21 640	21 640
	final segment	24 391	24 391
	TOR1	no limit	no limit
	TOR2	no limit	no limit
	TOD1	no limit	no limit
	TOD2	no limit	no limit
	ASD	no limit	no limit

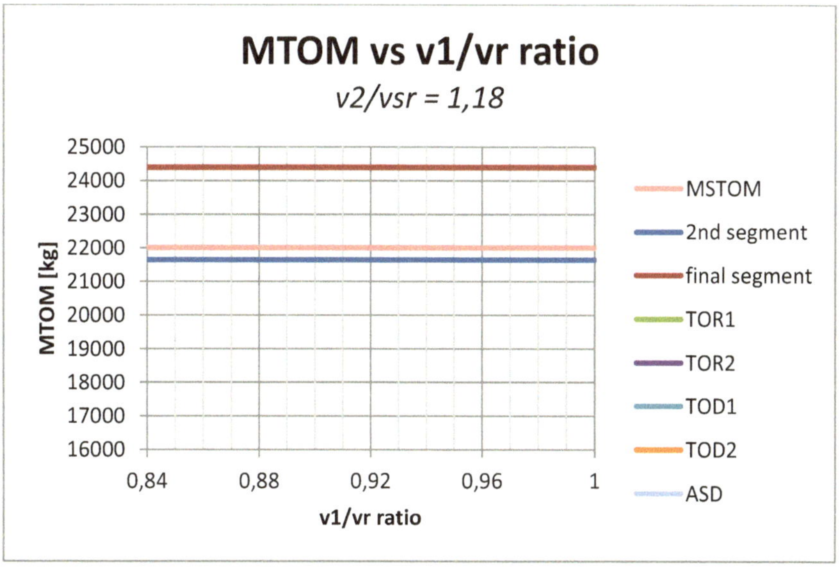

Figure A13: Rwy 01 MTOM vs. v_1/v_r ratio graph

LZKZ RWY 19

LZKZ RWY 19		v1/vr	
v2/vsr = 1,18		0,84	1
mass [kg]	MSTOM	22 000	22 000
	2nd segment	21 640	21 640
	final segment	24 391	24 391
	TOR1	no limit	no limit
	TOR2	no limit	no limit
	TOD1	no limit	no limit
	TOD2	no limit	no limit
	ASD	no limit	no limit

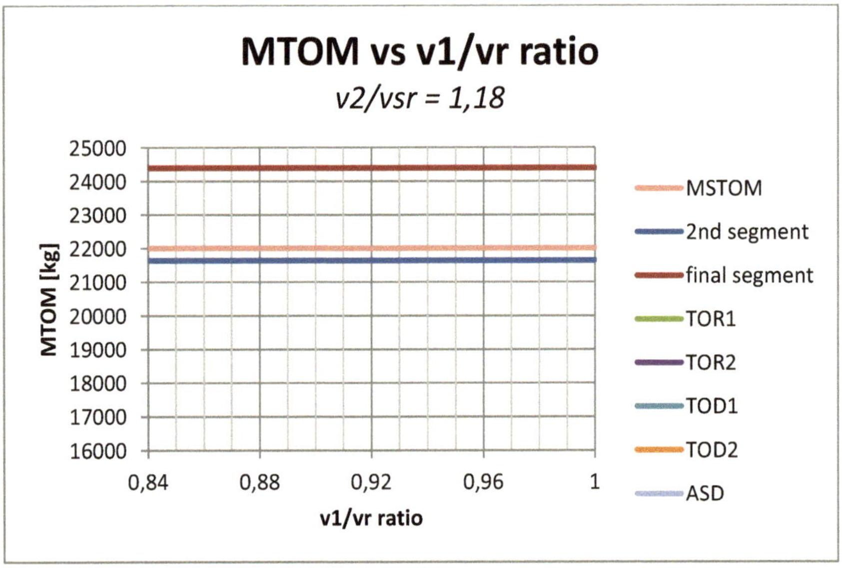

Figure A14: Rwy 19 MTOM vs. v_1/v_r ratio graph

Optimization process

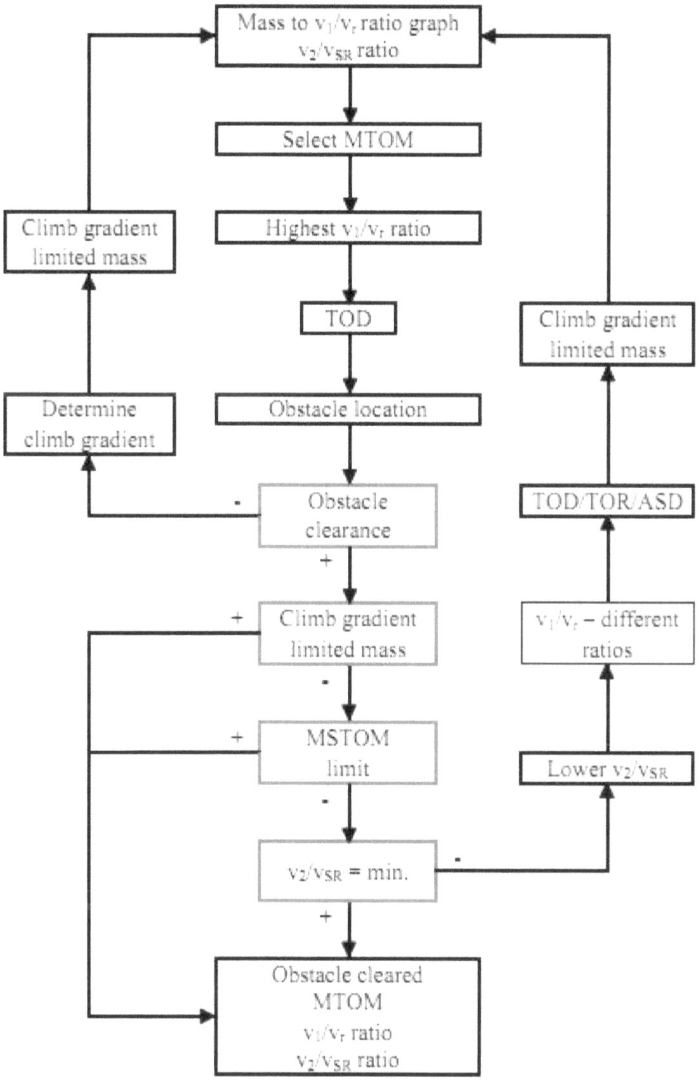

Figure A15: Optimization process flowchart

rwy 01	
MTOM	21 640 kg
v_1/v_r ratio	1
rwy 19	
MTOM	21 640 kg
v_1/v_r ratio	1

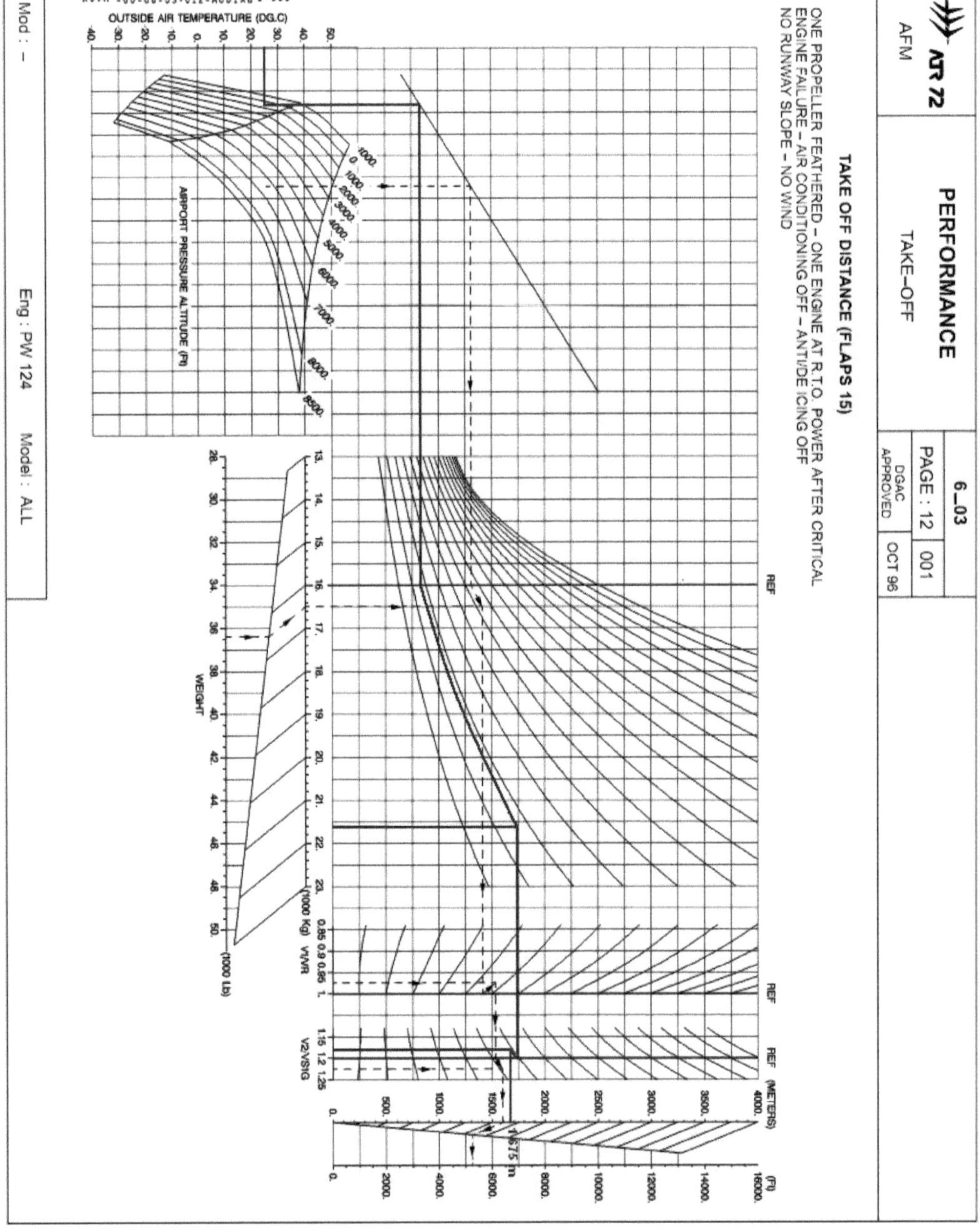

Figure A16: TOD rwy01/19 [Source: ATR, ATR 72 AFM]

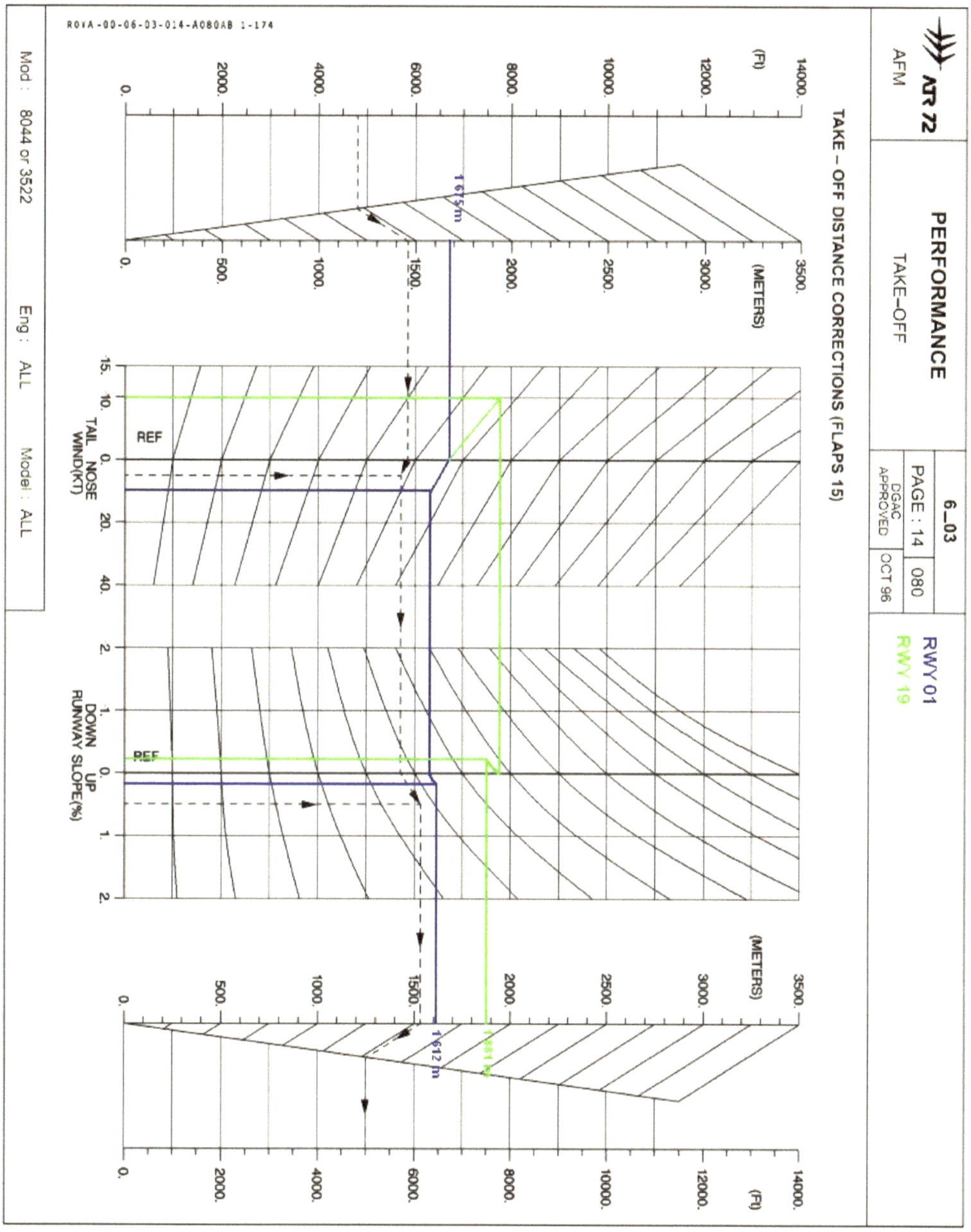

Figure A17: TOD rwy01/19 corrections [Source: ATR, ATR 72 AFM]

rwy 01	
TODA – line-up correction	3 289 m
TOD	1 612 m
Flight distance above rwy surface	1 677 m
rwy 19	
TODA – line-up correction	3 289 m
TOD	1 881 m
Flight distance above rwy surface	1 408 m

obstacle	Departure rwy	Flight distance	Obstacle elevation	Obstacle height AAL
A	**19**	29,1 km	1 245 ft	490 ft
B	**01**	9,9 km	1 253 ft	498 ft
C	**01**	10,8 km	1 348 ft	593 ft
D	**01**	11,9 km	1 610 ft	855 ft
E	**01**	15,6 km	1 784 ft	1 029 ft
F	**01**	16,6 km	1 847 ft	1 092 ft
G	**01**	24,1 km	2 368 ft	1 613 ft

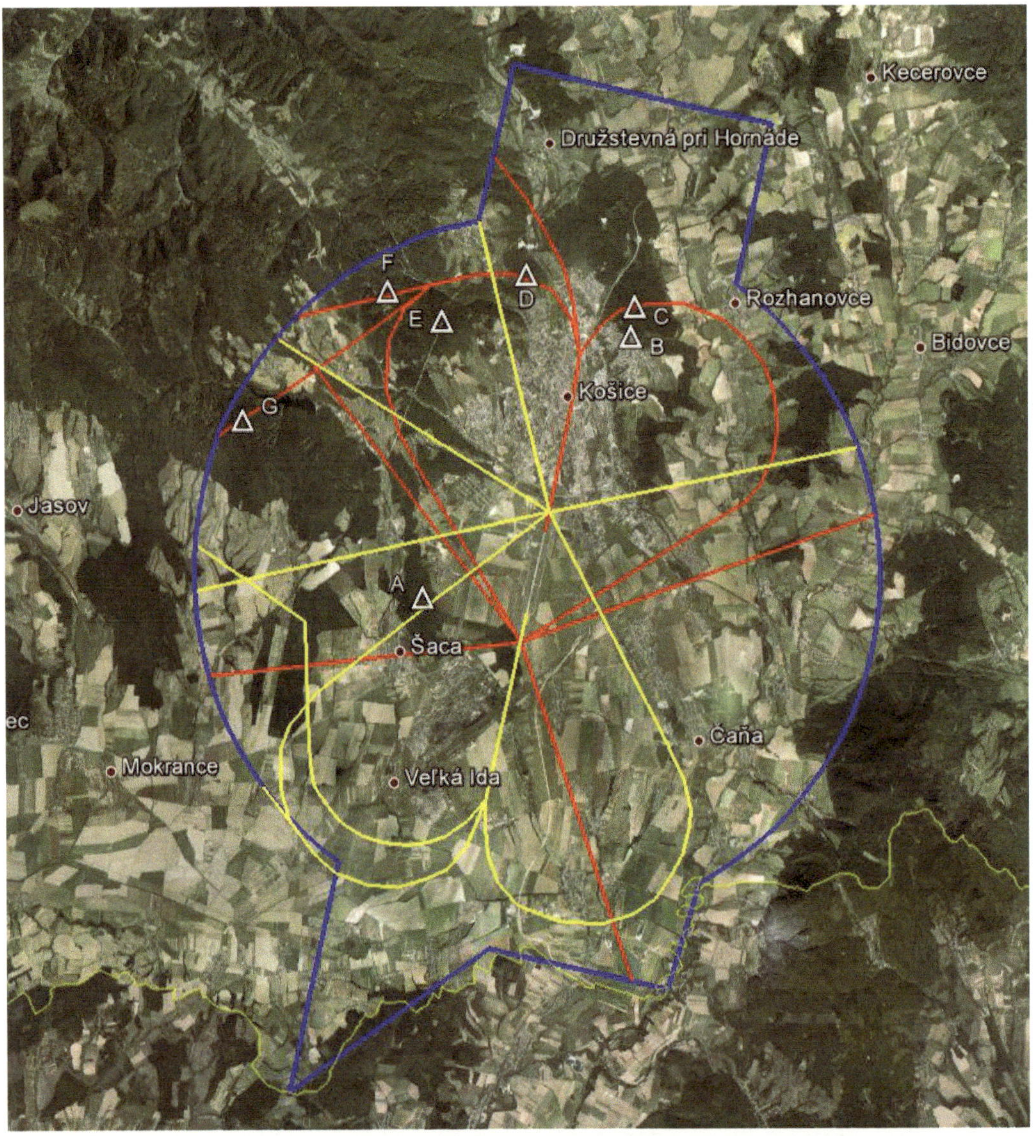

Figure A18: LZKZ obstacles

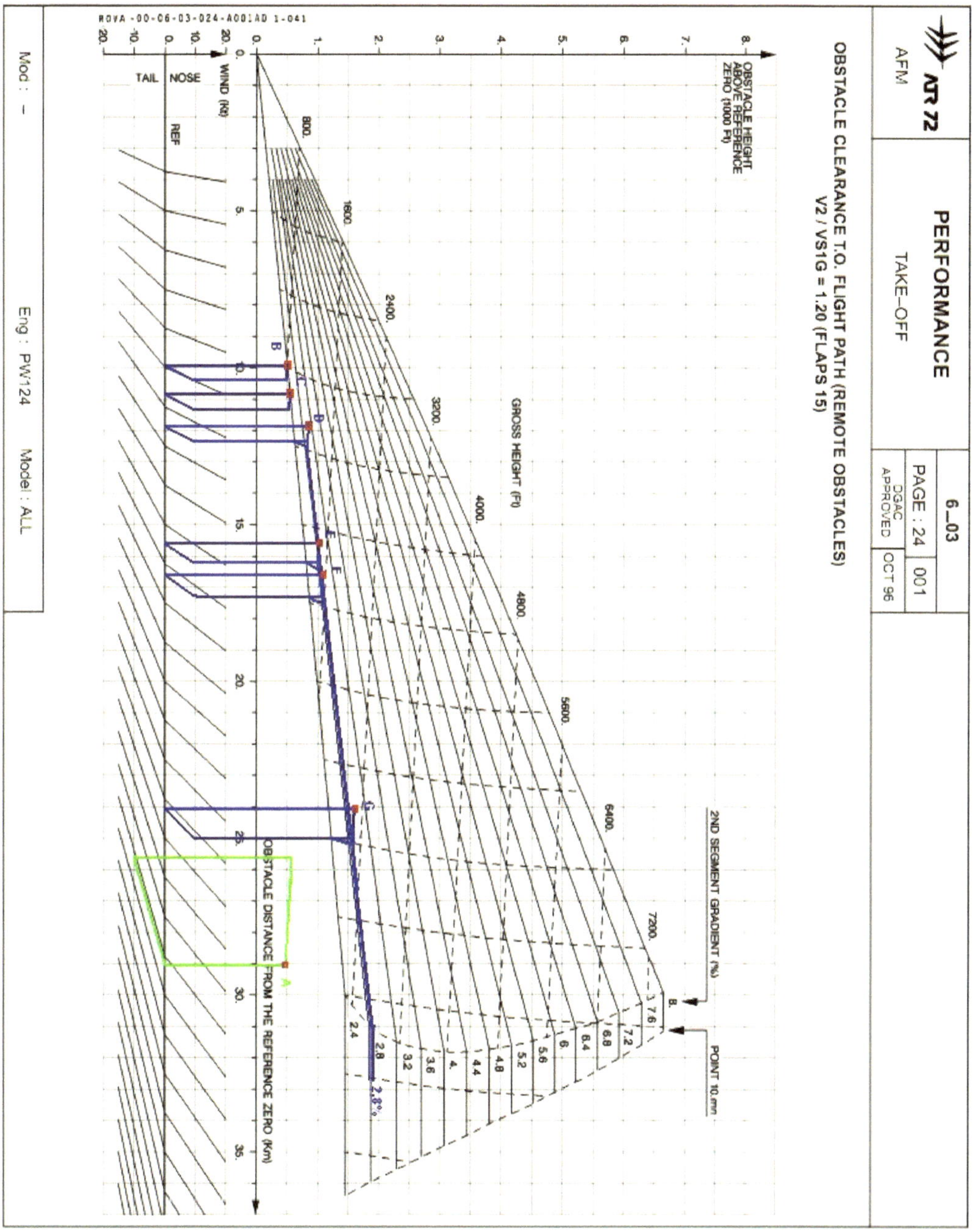

Figure A19: Obstacle clearance [Source: ATR, ATR 72 AFM]

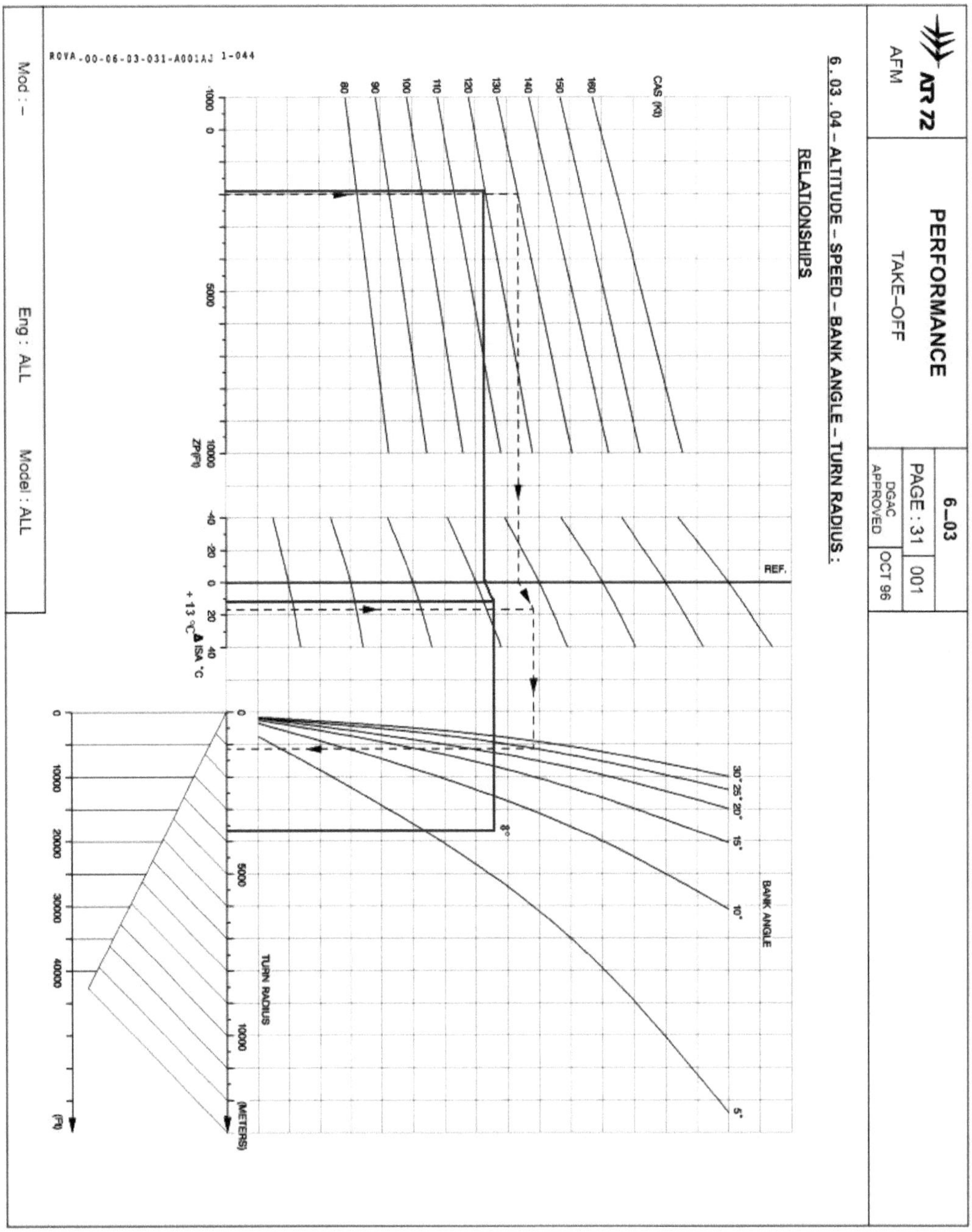

Figure A20: Turn radius [Source: ATR, ATR 72 AFM]

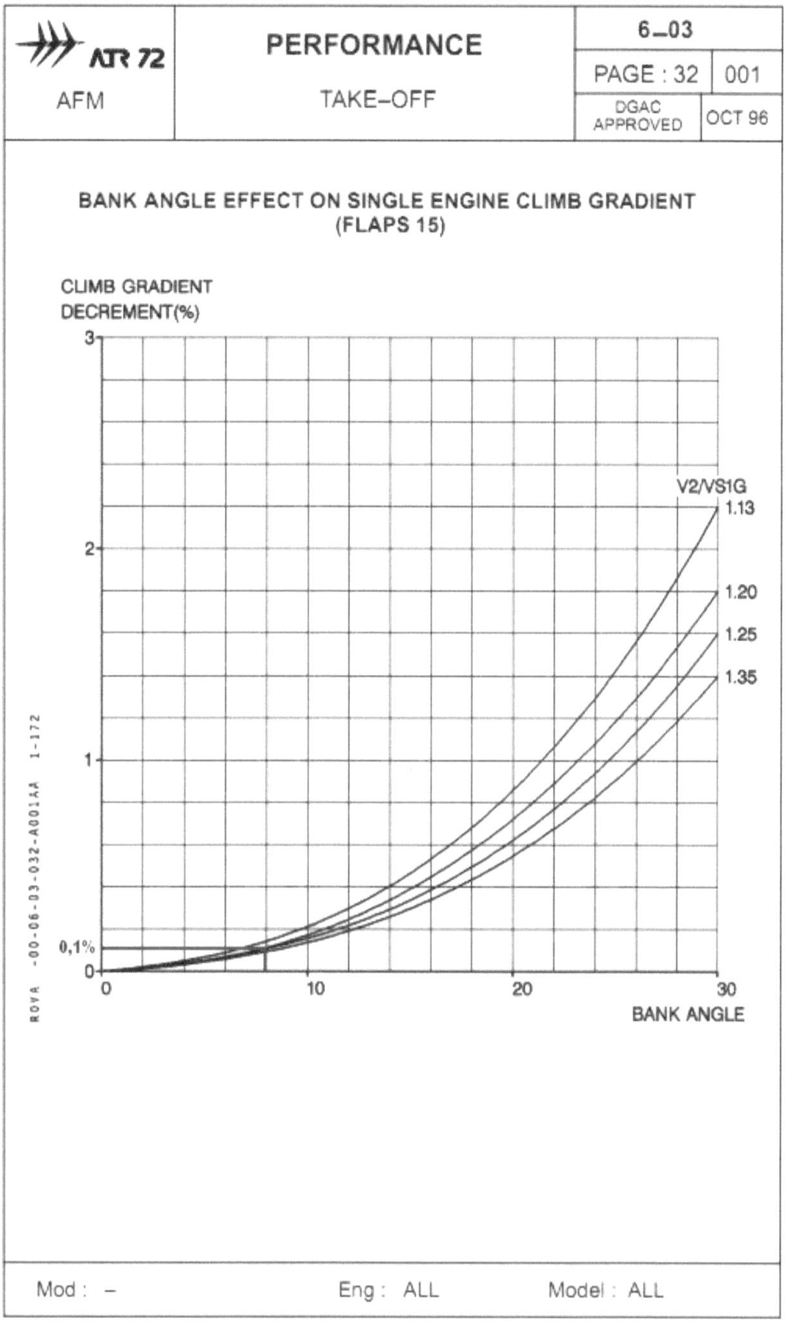

Figure A21: Bank angle c. grad. decrement [Source: ATR, ATR 72 AFM]

	Most critical obstacle	Obstacle climb gradient	Bank angle climb increment	Required climb gradient	Obstacle cleared
rwy 01	**D**	2,8%	0,1%	2,9%	NO
rwy 19	**A**	no limit	0,1%	no limit	YES

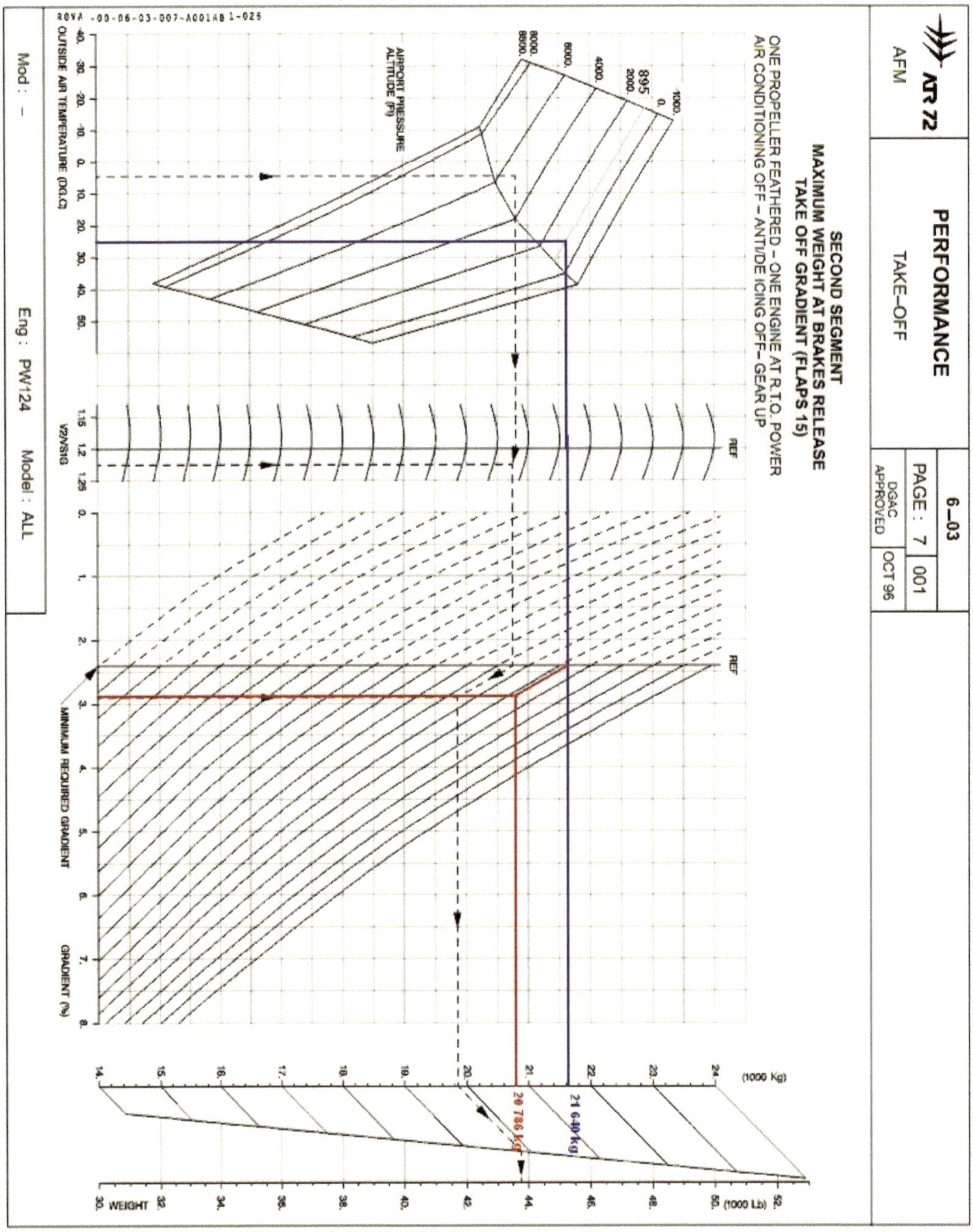

Figure A22: 2nd segment climb gradient adjusting [Source: ATR, ATR 72 AFM]

LZKZ RWY 01 obst. correct.		v1/vr	
v2/vsr = 1,18		0,84	1
mass [kg]	MSTOM	22 000	22 000
	2nd segment	20 786	20 786
	final segment	24 391	24 391
	TOR1	no limit	no limit
	TOR2	no limit	no limit
	TOD1	no limit	no limit
	TOD2	no limit	no limit
	ASD	no limit	no limit

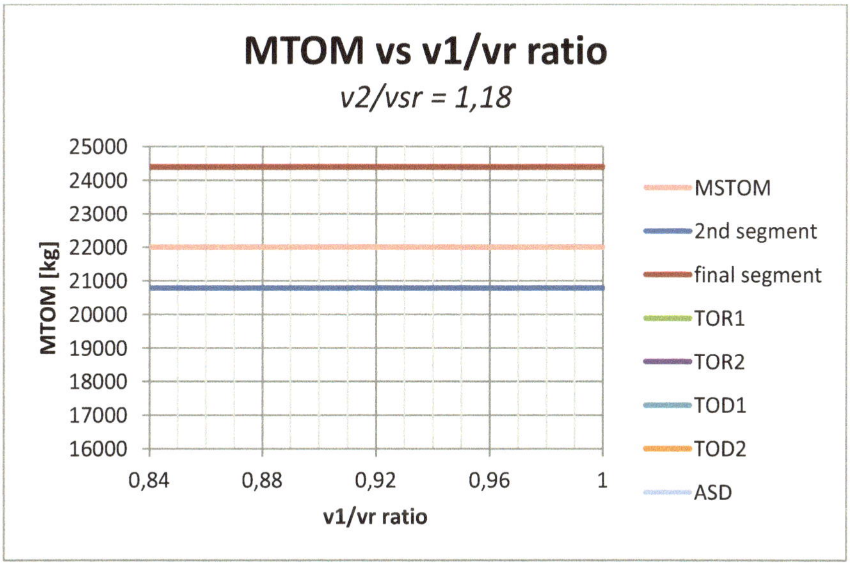

Figure A23: Rwy 01 MTOM vs. v_1/v_r ratio obstacle corrected graph

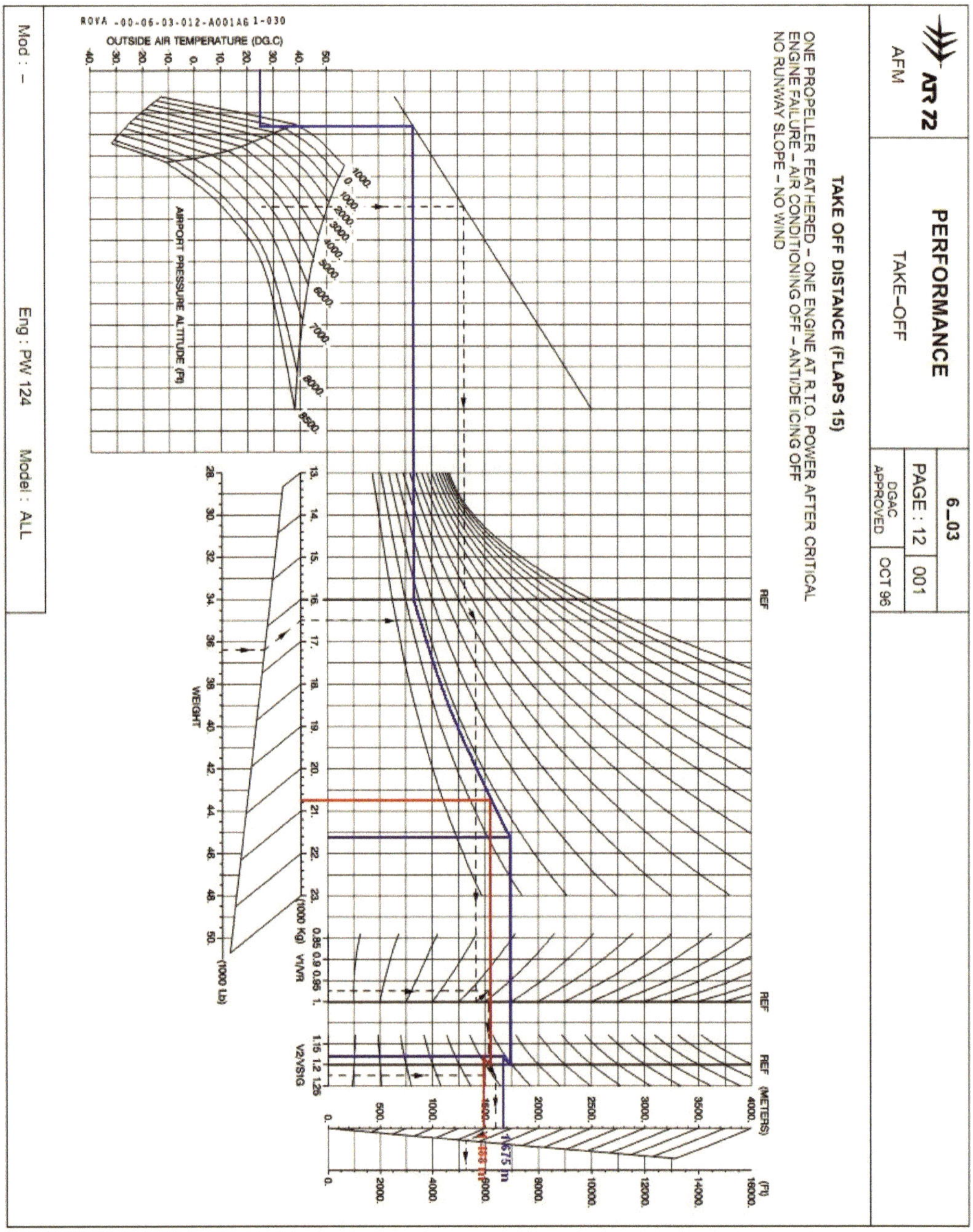

Figure A24: TOD rwy01 adjusting [Source: ATR, ATR 72 AFM]

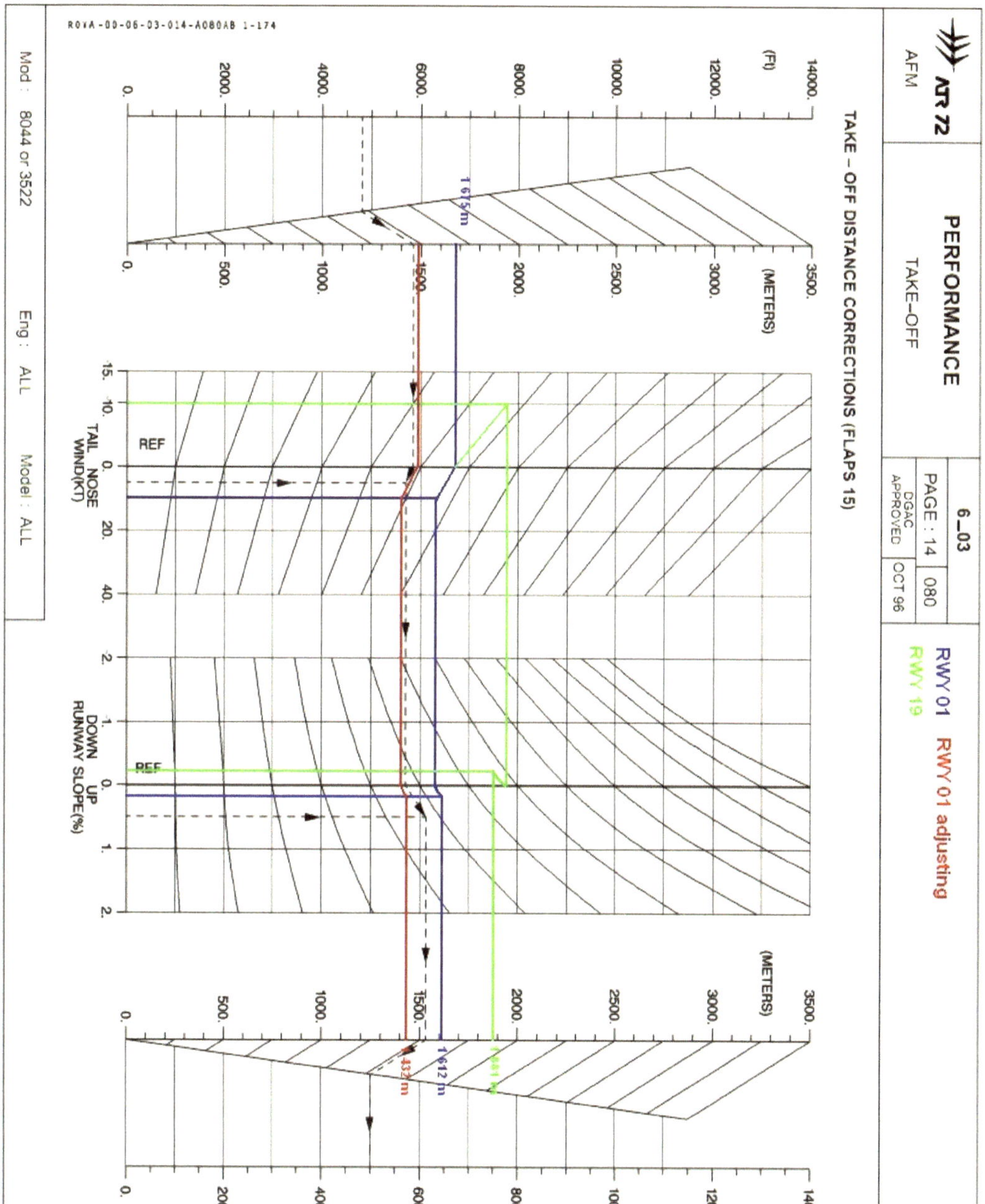

Figure A25: TOD corrections adjusting [Source: ATR, ATR 72 AFM]

rwy 01	
TODA – line-up correction	3 289 m
TOD	1 432 m
Flight distance above rwy surface	1 857 m
Obstacle D	
Flight distance	12,1 km
Obstacle elevation	1 610 ft
Obstacle height AAL	855 ft

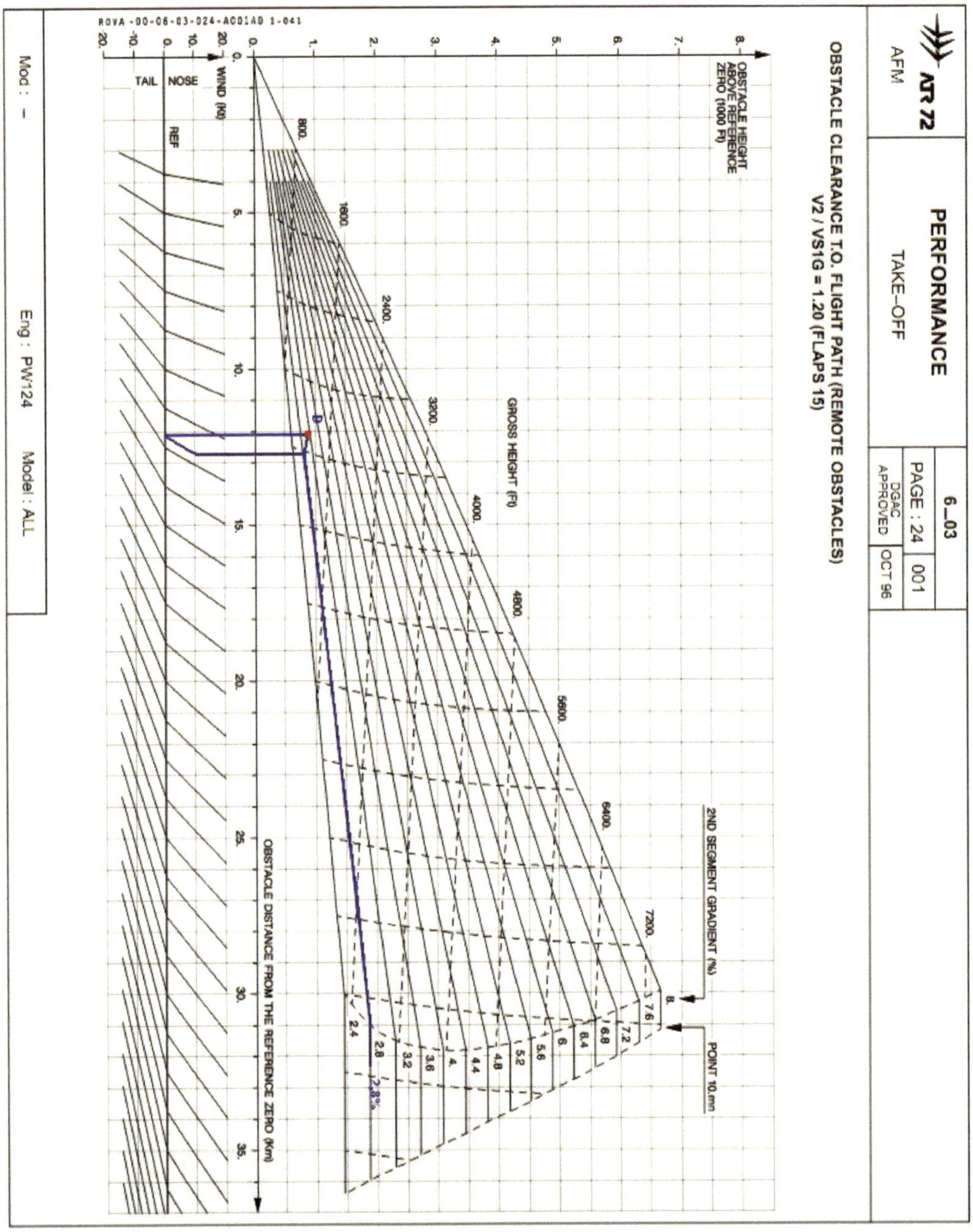

Figure A26: Obstacle clearance adjusting [Source: ATR, ATR 72 AFM]

	Most critical obstacle	Obstacle climb gradient	Bank angle climb increment	Required climb gradient	Obstacle cleared
rwy 01	**D**	2,8%	0,1%	2,9%	YES
rwy 19	**A**	2,2%	0,1%	2,3%	YES

rwy 01	
MTOM	20 786 kg
v_1/v_r ratio	1
v_2/v_{SR} ratio	1,18
Obstacle clearance	YES
rwy 19	
MTOM	21 640 kg
v_1/v_r ratio	1
v_2/v_{SR} ratio	1,18
Obstacle clearance	YES

v-speeds evaluation

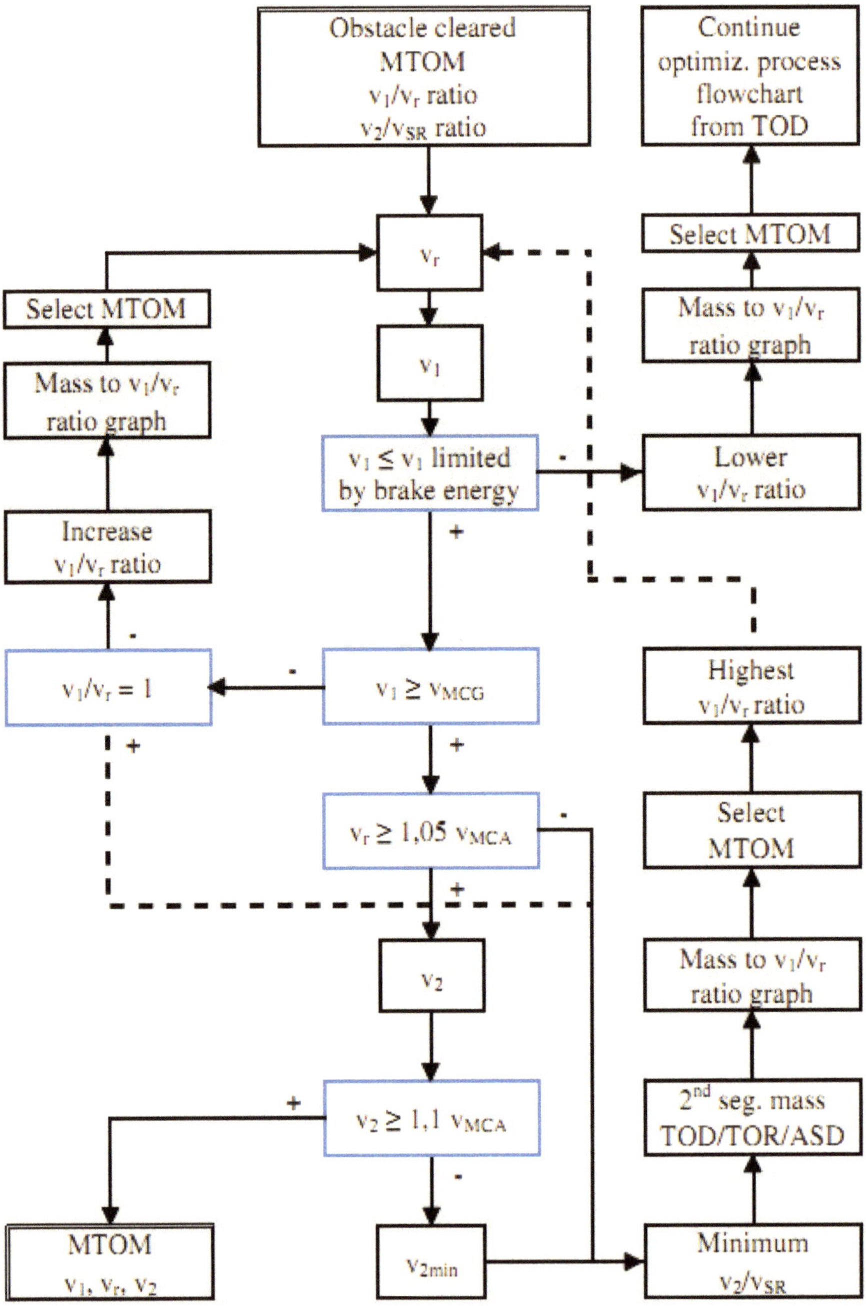

Figure A27: v-speeds evaluation flowchart

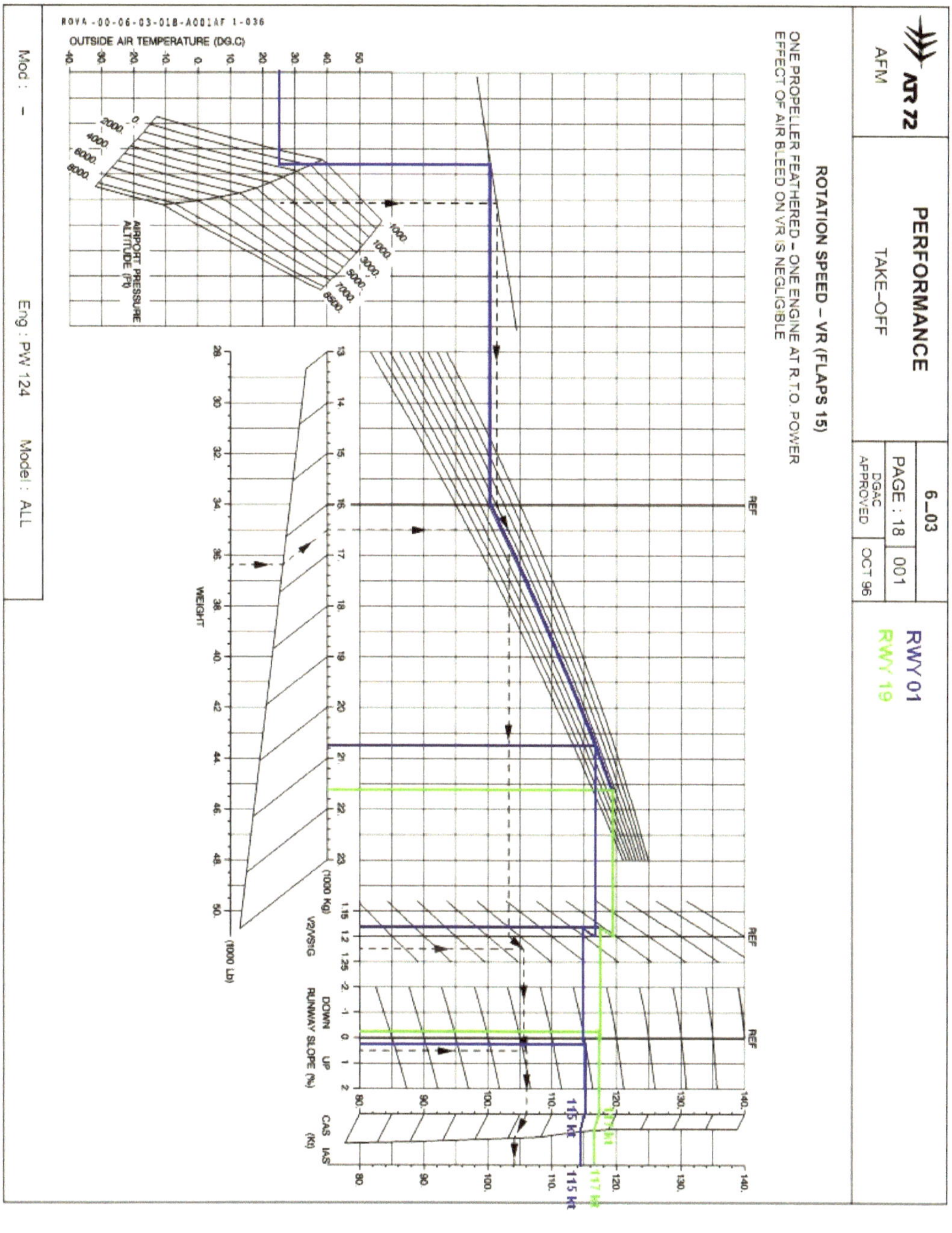

Figure A28: v_r evaluation [Source: ATR, ATR 72 AFM]

v_1/v_r ratio		1
rwy 01	**v_r**	115 kt
	v_1	115 kt
rwy 19	**v_r**	117 kt
	v_1	117 kt

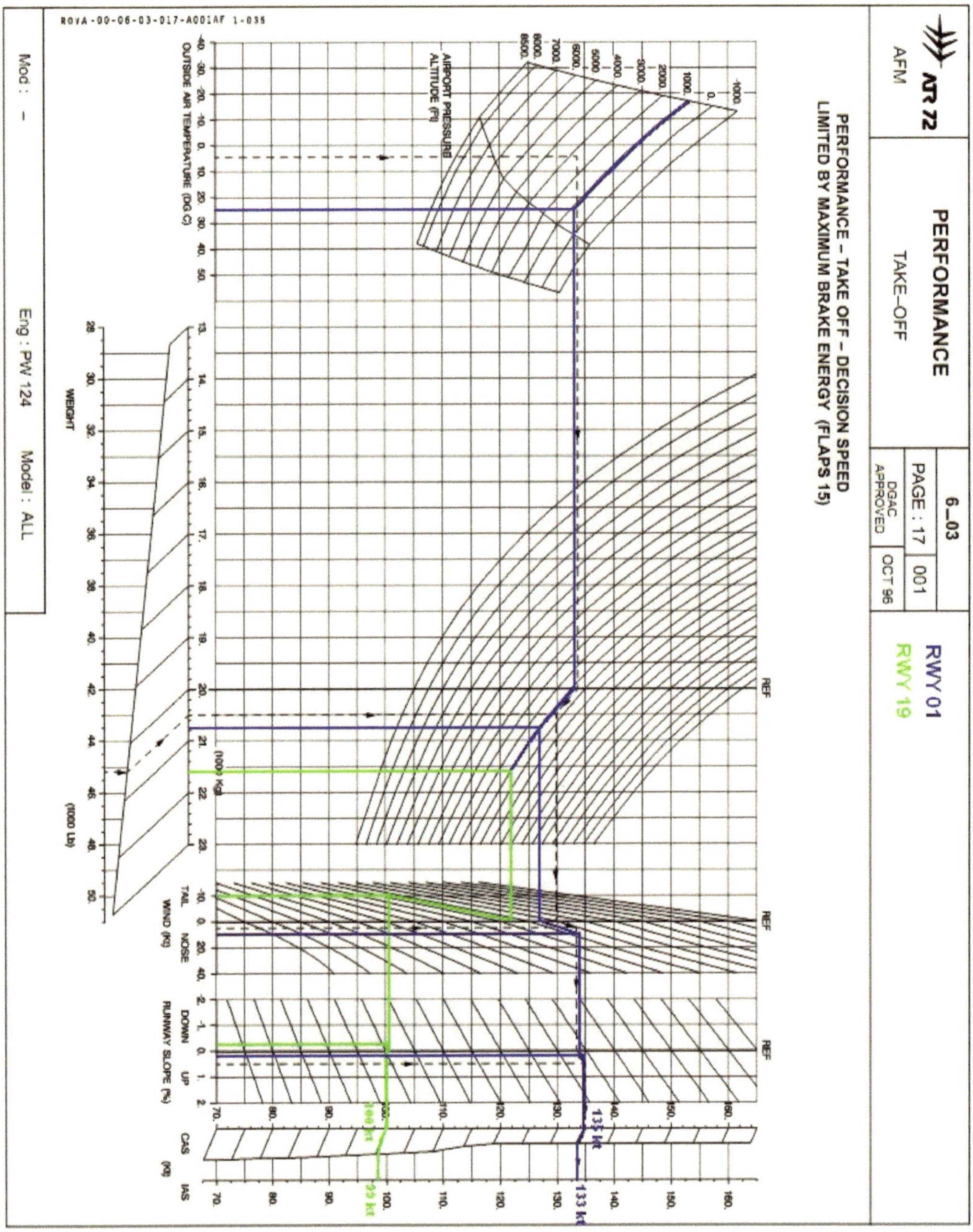

Figure A29: v_1 limited by max. brake energy [Source: ATR, ATR 72 AFM]

Calculated v_1 for the runway 19 (117 kt) does not comply with a maximum value of the v_1 limited by the brake energy (99 kt). Therefore, the v_1/v_r ratio needs to be decreased. By comparing appropriate values, this ratio should be set to a value of approximately 0,85. This provides with a new v_1 (99 kt). According to the v-speed evaluation flowchart, a new MTOM need to be selected for this ratio. Moreover, the TOD and obstacle clearance need to be checked again. As it could be observed from the "MTOM vs. v_1/v_r ratio" graph for the runway 19, this ratio change does not influence the MTOM value or its limitation. Consequently, after checking for obstacle clearance we may continue with already calculated value of v_r, as it is not changed.

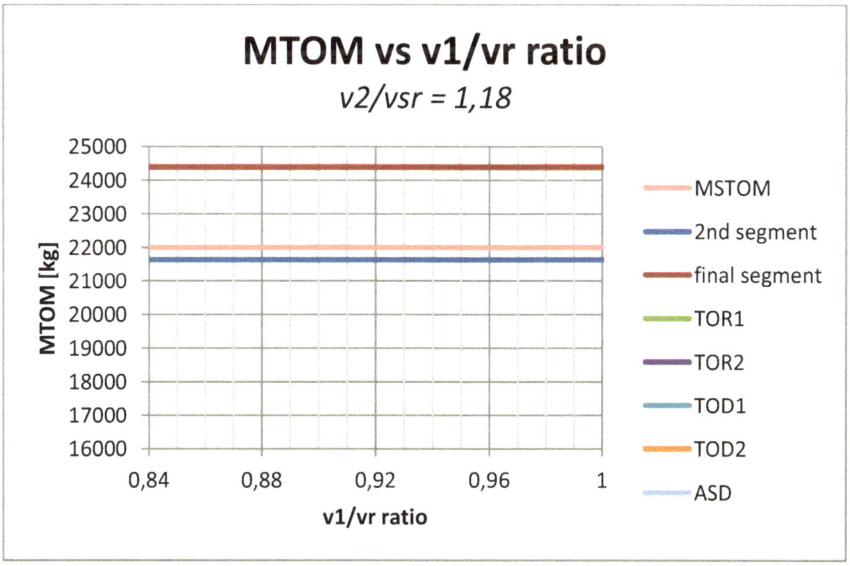

Figure A30: Rwy 19 MTOM vs. v_1/v_r ratio graph

rwy 01	**v_1/v_r ratio**	1
	v_r	115 kt
	v_1	115 kt
rwy 19	**v_1/v_r ratio**	0,85
	v_r	117 kt
	v_1	99 kt

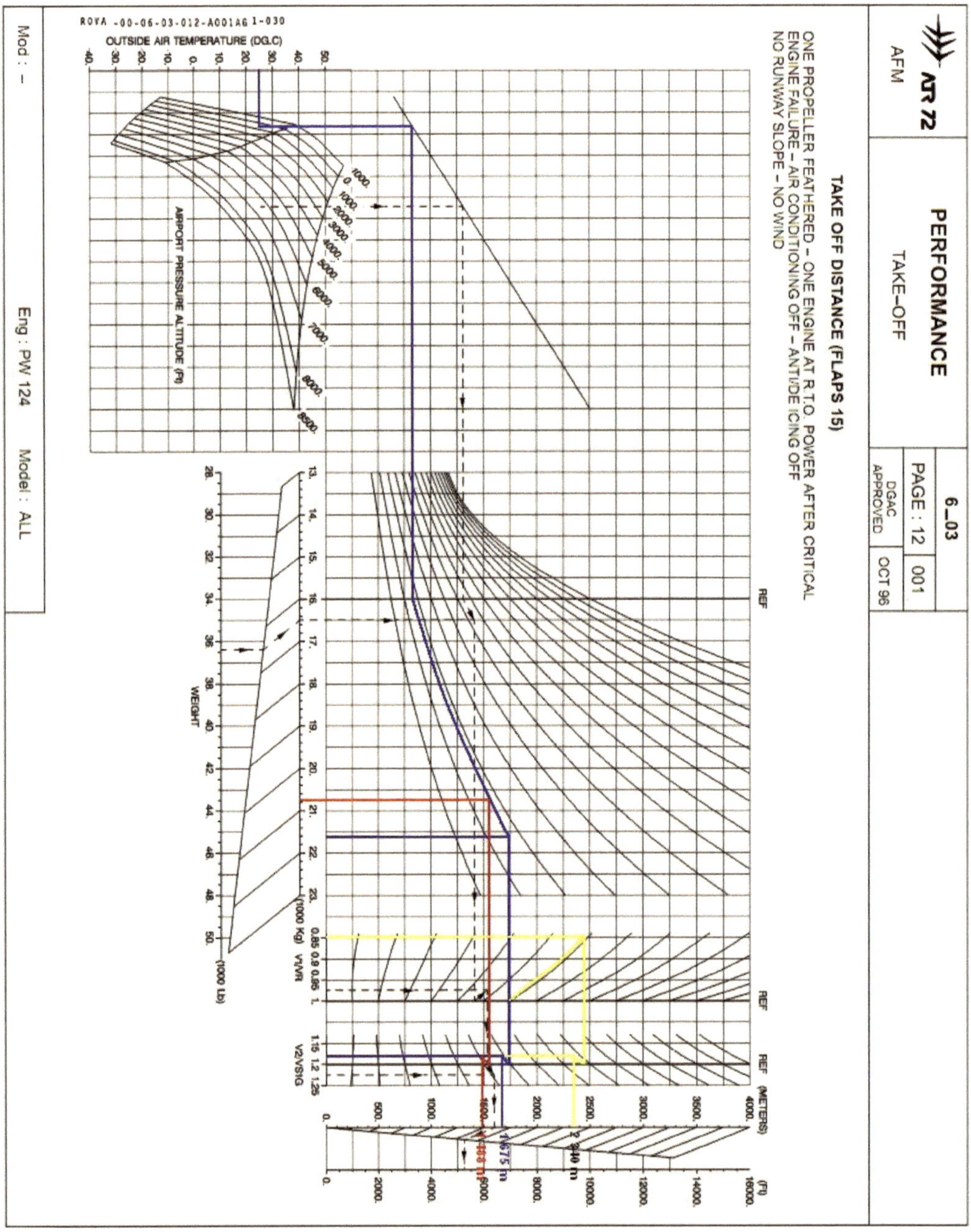

Figure A31: TOD rwy 19 adjusting [Source: ATR, ATR 72 AFM]

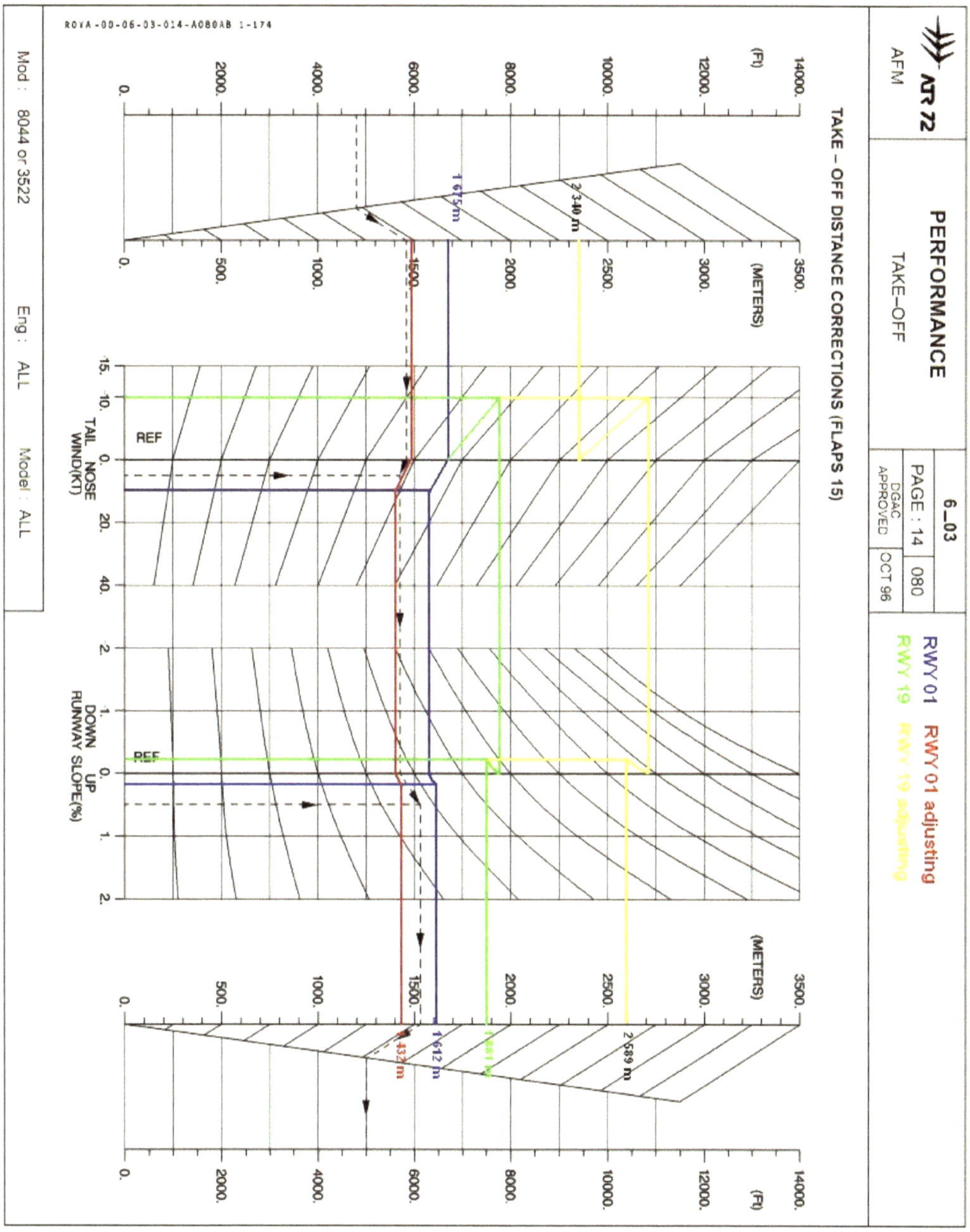

Figure A32: TOD corrections adjusting [Source: ATR, ATR 72 AFM]

rwy 19	
TODA – line-up correction	3 289 m
TOD	2 589 m
Flight distance above rwy surface	700 m
Obstacle A	
Flight distance	28 km
Obstacle elevation	1 245 ft
Obstacle height AAL	490 ft

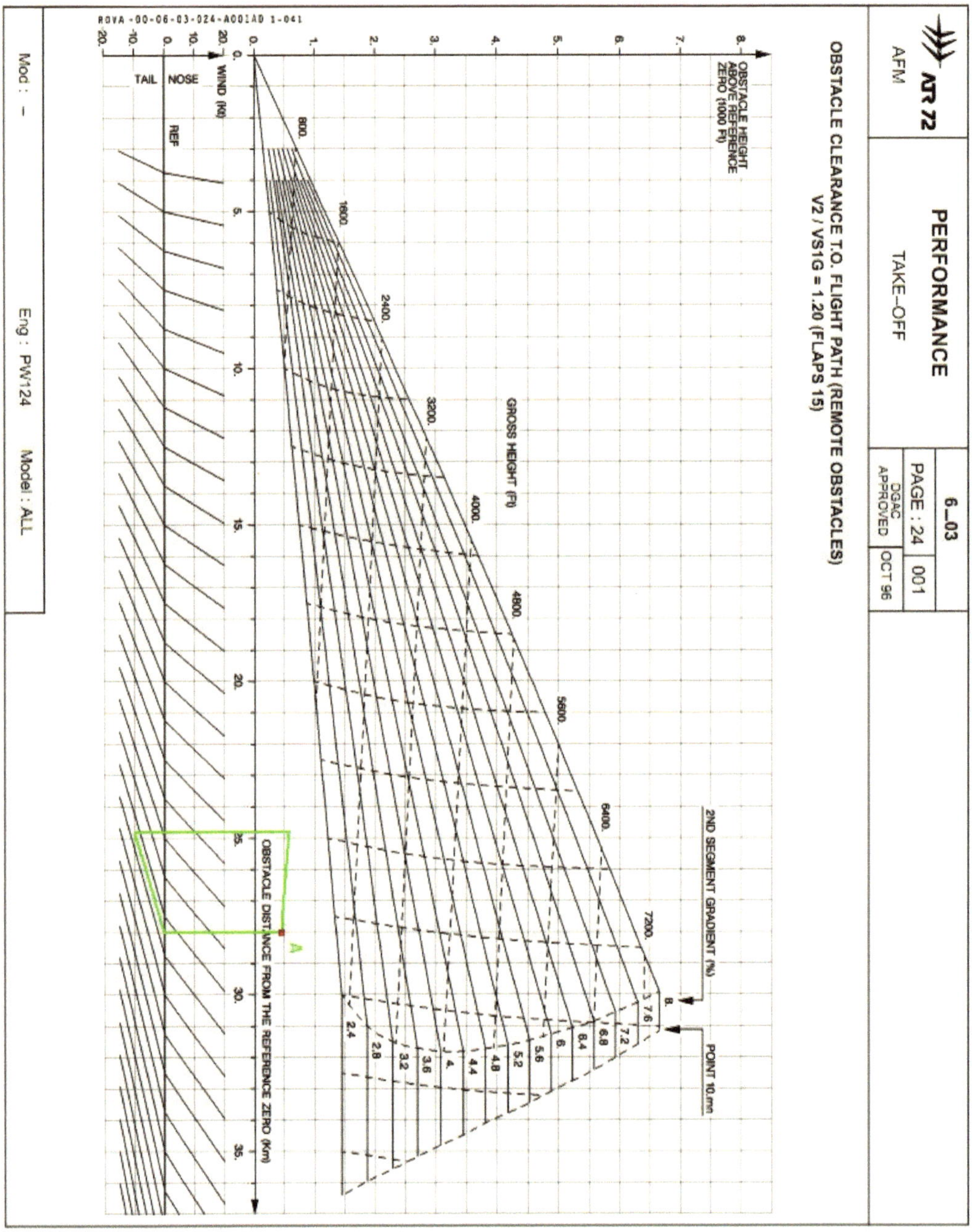

Figure A33: Obstacle clearance rwy19 adjusting [Source: ATR, ATR 72 AFM]

	Most critical obstacle	Obstacle climb gradient	Bank angle climb increment	Required climb gradient	Obstacle cleared
rwy 01	**D**	2,8%	0,1%	2,9%	YES
rwy 19	**A**	no limit	0,1%	no limit	YES

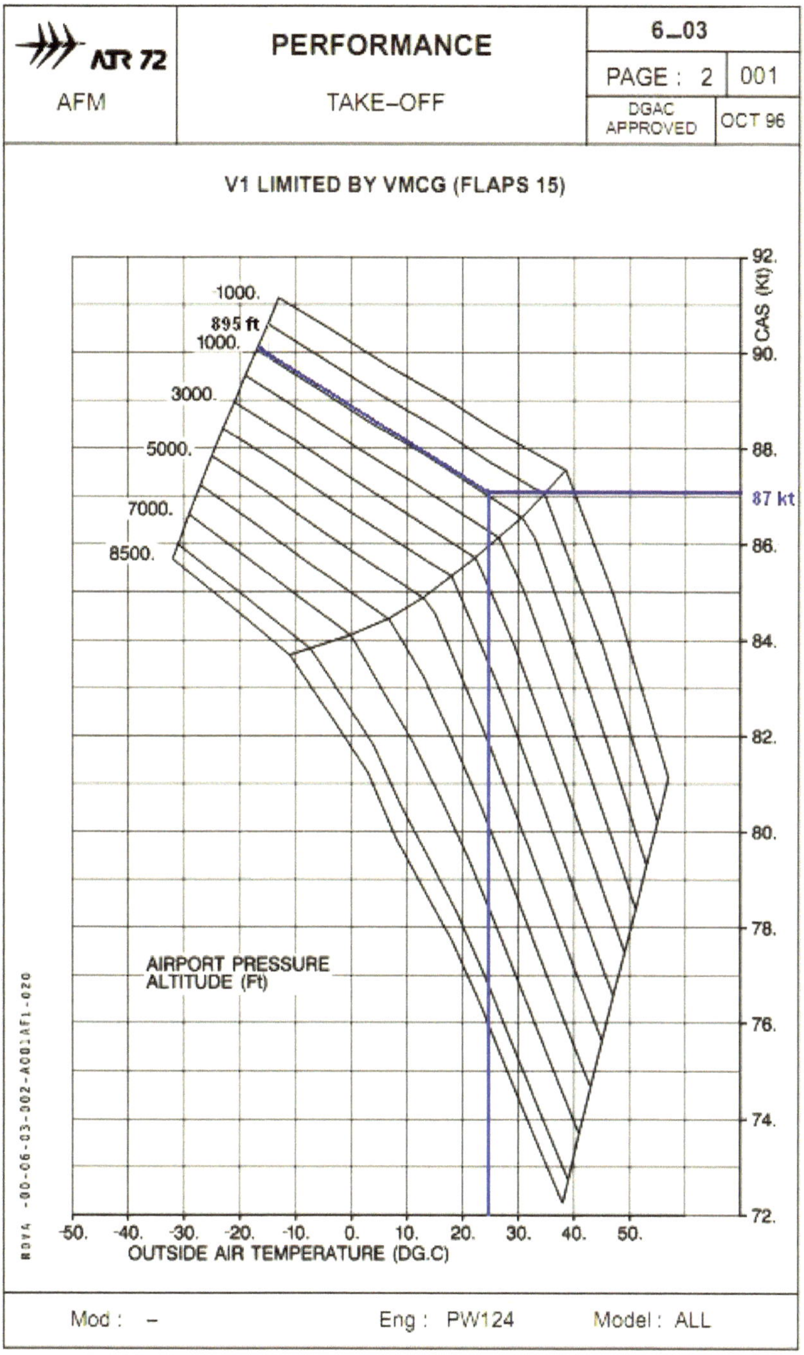

Figure A34: v_1 limited by v_{MCG} [Source: ATR, ATR 72 AFM]

	v_1	v_{MCG}
rwy 01	115 kt	87 kt
rwy 19	99 kt	87 kt

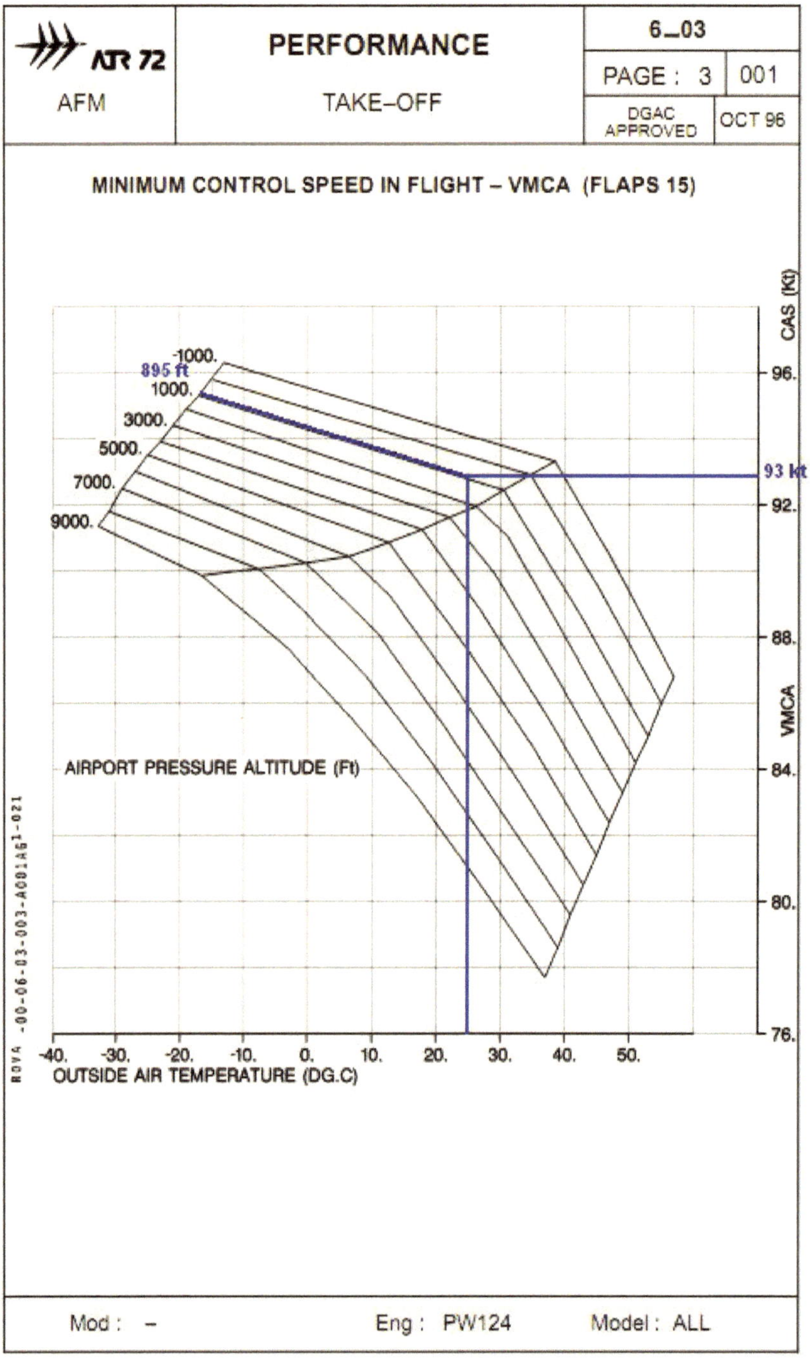

Figure A35: v_{MCA} [Source: ATR, ATR 72 AFM]

	v_r	**1,05 v_{MCA}**
rwy 01	115 kt	98 kt
rwy 19	117 kt	98 kt

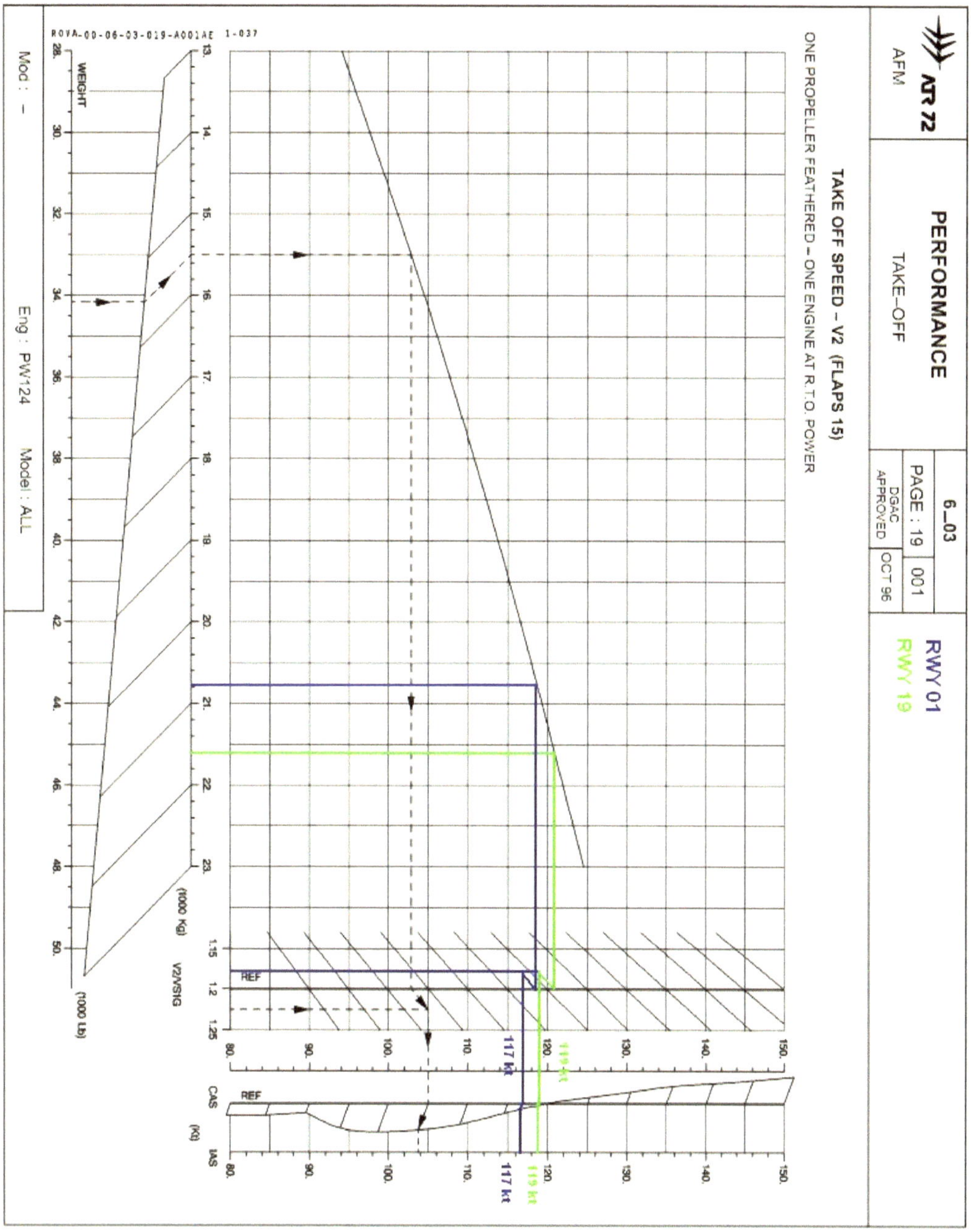

Figure A36: v_2 [Source: ATR, ATR 72 AFM]

	v_2	1,10 v_{MCA}
rwy 01	117 kt	102 kt
rwy 19	119 kt	102 kt

Conclusion

rwy 01	
MTOM	20 786 kg
v_1	115 kt
v_R	115 kt
v_2	117 kt
limitation	obstacle

rwy 19	
MTOM	21 640 kg
v_1	99 kt
v_R	117 kt
v_2	119 kt
limitation	maximum brake energy 2nd segment climb gradient

The tables above represent results of the take-off data calculation for the Košice airport and given conditions. Although the wind direction is more favorable for the take-off from runway 01, required obstacle clearance reduced the MTOM to a lower value than it is for runway 19. Therefore, to achieve the maximum allowable payload it is recommended to perform the take-off from runway 19 even despite current wind conditions.